SOY
SOURCE

SOY SOURCE

John Downes

PRISM PRESS

First published in Australasia in 1987 by
Nature & Health Books
This edition co-published in Great Britain by
Prism Press, 2 South Street, Bridport,
Dorset DT6 3NQ, England
and distributed in the United States of America by
the Avery Publishing Group Inc.,
350 Thorens Avenue, Garden City Park, New York, 11040

ISBN 1 85327 00 1 6

616.5

Series editor: Nevill Drury
Design: Craig Peterson

Typeset by Setila Type Studio, Sydney
Printed in Australia by The Book Printer

CONTENTS

This book is dedicated to Maar, Jesse, Sam and Nancy, with love.

I would like to express my gratitude to Nevill at Nature and Health, David for arranging it all, Christobel (alias Garimo) for enduring support, Sandra for processing the words so well, Maar as the source of it, and family, friends and fellow workers for tolerating me! And thank you to those crazy gods for the soybean.

INTRODUCTION

The many foods which are naturally produced from the soybean are a wonderland for anybody interested in cooking. Even more so because soyfoods meld with our established Western cuisine, considerably expanding our options. It is no secret that the large amounts of animal foods, fat and white sugar, which in one way or another form the basis of modern Western cooking, are, among other factors, instrumental in our rising tide of degenerative disease. Soyfoods provide a medium through which to reduce or even eliminate this heavy reliance on potentially unhealthy ingredients. Soyfoods are generally high in easily assimilable protein, low in fat, with no cholesterol, and are extracted from an ecologically sound vegetable source.

Such dishes as tempeh pizza, tofu mousseline and soymilk custard are simple ways in which a conventional Western dish can be transformed. Soyfoods simply open a whole avenue of creative endeavour in cooking and, using the traditional Asian methods of preparing them, lend an exotic and enlivening edge to your daily eating. You can be sure the dishes you prepare are nutritionally sound, with delicious flavour and flair.

In the 1950s classrooms, where children were pondering the future, space exploration loomed large and so did future foods. The image I had was of dried, brown, biscuity chunks which weren't much to eat but were magically sustaining. Yes, they were made from soybeans. The reality of soyfoods, as a gift from the East, couldn't be more different.

From its origins in the mists of time, the soybean emerged as a food crop in China over 4000 years ago. Its cultivation rapidly spread to Japan and South-East Asia, and it has been a staple of diet in these areas ever since. It is the perfect dietary complement to a people whose basic staple is rice and other cereal grains.

Even well-cooked whole soybeans contain an anti-nutritive substance called the *trypsin inhibitor*. This effectively stops us from secreting the necessary factors to digest the protein. It is fascinating that in those countries where soybeans are traditional fare, the only time the locals eat them whole is before harvest, as delicious green beans. Otherwise they are naturally processed in a variety of ways to maximise their protein availability and to render them digestible. We call these products *soyfoods*.

Because soybeans are of Asian origin, soyfoods have varying names, depending on the country in which they are used. Each country also has its own varieties and method of making soy products, so we have items as varied as rich, dark *miso paste*, and beans covered in a soft, white Camembert- like mould — *tempeh*.

The soyfoods which I have chosen to use in this book are :

Tofu or Beancurd	A soft white curd which when pressed resembles fresh cheese.
Miso	A paste of fermented soybeans, cereal grain and sea salt.
Soy sauce	(1) Shoyu: A dark sauce made by fermenting soybeans and wheat in water. (2) Tamari: Similar to shoyu, but contains no wheat.
Soy milk	A nutty milk extracted from soybeans with water. This is curdled to make tofu.
Tempeh	Fermented soybeans in a block, bound together by a white mould or mycelium — much like soft ripened cheese (e.g. Camembert).
Yuba	Soybean skin (or sheets) — skimmed from soymilk like scalded cream.

These foods all have non-English names, of course, and some just do not translate! The rich aroma of good miso, its envigorating quality, variety of colours and tastes, its honoured place in Japanese culture, are not at all translated by the term 'fermented soybean paste' or 'soybean purée'. In the same way, we prefer to call curdled, fermented, aged cow's milk 'cheese'.

We are just beginning to discover what Asian cultures have understood for aeons. Many authorities today are urging us all to eat less dairy products and red meat, and soyfoods have many answers . . . low fat, no cholesterol, fibre, adequate protein (which is increased considerably if eaten with whole cereal grains, the most recommended food group), and last but not least, interesting and enjoyable food. To top this off, we have a wealth of cooking tradition with soyfoods from which to glean inspiration.

Tofu

When soaked soybeans are ground with water, cooked and strained, a 'milk' known as soymilk is formed. If this milk is curdled and the resultant whey separated from the curds, we have in its most basic form, soybean curds or, in

Japanese, tofu. These curds are pressed under weights between cloths in settling boxes and the resultant product can then be cut into gleaming, creamy-white cakes which may be familiar to you from Chinese food stores, or more recently, from natural food shops.

The Chinese call it *dofu or dow-fu;* the Indonesians *tahu,* the Burmese *tohu* or *pepya,* and generally in Thailand, Laos and Vietnam, it follows the Chinese pronunciation. I will call it tofu throughout this book as it is becoming widely known by this name, especially as a natural food. Those introduced to it by the Chinese will know it as bean curd.

Tofu originated in China over 2000 years ago and it is a familiar ingredient in Chinese cooking today, indeed forming the protein mainstay of the Chinese diet. It rapidly spread to South-East Asia and Japan, where it also figures significantly in the cooking of these nations. Tofu has become localised in these countries and each has numerous methods of preparing it, much to the delight of anyone interested in cooking or eating.

From a culinary standpoint, tofu is very exciting. Not only does it open up an avenue of exploration into the exotic cooking of South and East Asia, the cooking you will enjoy will not be laden with fats and sugar. In short, forget about being 'health conscious' – you *know* you will be eating high-quality, life-sustaining food. Tofu is relatively bland tasting, making it a good counterpoise to delicious sauces. Its texture is soft and creamy, a delight to contrast with crisp and hard textures, or to use to full advantage in dips or generally as a replacement for many dairy items.

There are a number of varieties of tofu. Those generally available are:
1. hard,
2. medium soft, and
3. soft.

Hard tofu is made by pressing the curds more than for regular tofu. This gives a compact texture which will not disintegrate when being stir-fried, for example. Hard tofu is generally available in Chinese stores, and is quite distinguishable, being thinner (from greater compression) and obviously more solid than their regular tofu.

Medium-soft tofu is a catchall which describes the Chinese product and some of the newer natural tofu. The Chinese one is generally softer than the natural food store product – sometimes startling so. Chinese tofu can be quite jelly-like, although still firm. The burgeoning soyfoods market in the USA demanded tofu with a texture which is suitable for incorporation into Western-style cooking, as well as for Asian styles. The new 'natural tofu' makers in Australia have generally taken their lead from the American adventure, hence this 'new style' tofu.

Soft tofu is called *kinugoshi* or *'silken'* in Japan, and *tahu fah* throughout South-East Asia. This is not pressed in the production, the curds being allowed to mingle and set in the whey. Whey is then carefully scooped off. This tofu has a delicate texture which really is silken and glistening. In Japan, the texture is prized as the basis of beautiful arrangements. I have used it as a garnish, particularly atop dishes which have dark and bright colours – on a whole baked fish glazed with black bean sauce it was very effective.

In South-East Asia, silken tofu can be bought still warm from many a street vendor, who serves it with all manner of side foods such as ginger tea, tapioca, agar, puffed barley, deep-fried gluten puffs, lotus seeds and natural black or palm sugar. One tahu-fah pedlar we frequented in Penang scooped the delicate curds out of his scrubbed wooden vat with a beautiful sea shell. These scoops were then laid on top of each other, black sugar syrup added, and it was slurped down with quiet satisfaction. We found tahu-fah to be a great boon when travelling with children, as it was widely available, fresh and nutritious. It was also very cooling in the tropical heat and, being such a light food, harmonised perfectly with the surroundings.

In Australia, tahu-fah is available in a few Chinese restaurants and from some Chinese bean-curd makers, if you visit the plant. Occasionally I have seen it in a plastic tub in the refrigerator of a Chinese food store. It can also be purchased in a long-life tetrapak under the brand name Morinaga Silken Tofu. This is a good product, but I use it only occasionally because I don't like the curdling agent used, which is GDL (Glucono-Delta-Lactone). Silken tofu can be made at home relatively easily (see p.37).

As I have mentioned, tofu is made by curdling soymilk. The agents used to curdle soymilk can vary and hence the quality of your tofu varies. *Nigari* is the traditional Japanese coagulant. It is the grey liquid which drips from a hessian bag of raw sea salt, and is mineral rich. Making good tofu with nigari is quite an art, as there are many subtle variables. However, it produces very good-tasting tofu.

The other coagulant generally used is *calcium sulphate*. Some purists think this is an artificial chemical. This is not true, and, in fact, calcium sulphate was the original and is probably the most natural curdling agent. That is, of course, if you get if from the right source. Calcium sulphate occurs naturally as gypsum and is abundant in Australia. It increases the percentage of calcium in the tofu and is easier to make softer tofu with. Some Chinese manufacturers use poor-quality calcium sulphate, which is an industrial by-product. Consequently, their tofu is less wholesome. I always buy tofu made from organic soybeans, and I make a point of buying it on the day it is delivered.,

I'm waiting for the day when one of our natural tofu makers use good-quality gypsum. It produces more nutritious tofu, with a very good texture, and it requires less skill. Nigari tofu is, to my mind, the best, but only in the hands of an experienced soycrafter.

There are a number of ways to buy tofu. Some Chinese will cut whatever size you please from large slabs laid out on stainless steel. In oriental food stores, it is packed in rectangular plastic containers and displayed in the refrigerator. Natural food stores and 'health' food stores carry the Chinese bean curd in the same containers and the natural products either vacuum packed or water packed in a sealed plastic container.

One of the problems some of our soycrafters experience is of maintaining deliveries of fresh tofu. A financially viable run of tofu-making can be done if a portion is vacuum packed. This vacuum-packed tofu is given a very long shelf life, and is sometimes quite sour. Soycrafters need to take great care with hygiene and cleanliness. The difference between fresh tofu and 6-week old, vacuum-packed tofu is dramatic — so support those who sell you the freshest tofu, and complain to the makers if yours is sour.

After buying tofu, unless you intend to use it immediately, store in the refrigerator. If you have firm tofu, cover it with water in a sealed container and refrigerate. Change the water every two days and it will keep well for a week. If you have no refrigerator, store in the coolest place in salted water (brine).

Agé-Tofu

Agé, dow-foo bok, fried bean curd is made by deep frying strips or small cubes of tofu. After frying, these have a tremendous texture and great possibilities. When bean curd is fried, a thick skin is formed and the interior becomes hollow or honeycombed. This can be sliced and added to stir fries, salads, cold dishes, noodles, etc., or prised open and stuffed with innumerable fillings. To remove most of the frying oil, agé is soaked in hot water briefly before use.

Agé is easy to make at home and can be refrigerated or frozen. It is well liked for its 'meaty' texture. The protein is concentrated by frying, hence agé is, by weight, very high in protein. Agé is always available in Japanese food stores where it comes in large and medium rectangles. The large pieces are excellent for stuffing, as they are big enough to form a neat, firm envelope, the top of which can be sealed and dropped in a simmering stock. The Chinese also make much smaller cubes or triangles of fried bean curd. These are either sliced and used in other dishes, or stuffed and served hot or cold.

Soymilk

The soybean earns its reputation as 'the vegetable cow'. Apart from producing cheese-like tofu, the most simple method of rendering the soybean into food is as soymilk. Soaked beans are ground with water. This slurry is then cooked and strained to yield a rich, creamy liquid. Soymilk has certainly been a phenomenon in Australia. It has always been available in Chinese food stores, loaded with sugar. The soy 'revolution' in the USA inspired several people in Australia, and home-made soymilk started to appear. It was delicious, especially if one ate no dairy foods, but many commented on its 'beany' flavour. In the ensuing years, after much experimentation, it has become clear that these beany flavours can be almost eliminated through careful preparation, even in the home kitchen (see Making Soymilk, p.36).

In the late 1970s a Japanese soymilk in plastic pouches appeared on the market. This was remarkable – it had no beany flavours and was very much like cow's milk in texture.

There are also another seven or eight soymilks on the market in tetrapaks today, some Japanese and a few Australian. These rate from terrible to fair. My personal preference is for fresh, home-made soymilk. The Japanese milk is very handy, and less of a taste challenge than the home-made one, but it is a highly processed food, and imported at that! The new Australian tetrapaks aren't much cheaper and they certainly don't compete on flavour or texture.

Bonsoy has the simplest ingredients – soybeans, pearl barley, kombu seaweed and water. Others, such as Pure Harvest, contain oils, salt and malt. Westbrae market an extraordinary range of 'milkshake' – like soy drinks in flavours such as carob malted, Java, carob mint. They are guaranteed to satisfy the wildest cravings! All of the above products are 'natural' to the extent they do not contain any chemical (artificial) ingredients. However, the newest soy drink to come on to the market is Sanitarium's So-good. It is a dreadful product, full of sugar and chemicals. It contains 'soy-isolate' rather than any natural soymilk, and is the sort of pseudo-food that should be avoided.

Experiment for yourself. Make some soymilk at home – a couple of times, until you've got the hang of it. Then compare all the commercial products and make your own choice.

Unfortunately, soymilk has had to run the gamut of seemingly endless and equally meaningless labelling requirements. At this point in time, it cannot legally be called soy 'milk'. Do you suppose somebody would mistake it for cow's milk? This is obviously a misguided attempt by our ever-vigilant bureaucracy to protect our dairy industry. (Why does it need protecting?) And to protect us! Perhaps mother's milk should be called mother's 'beverage', or 'drink'!

One of the great boons of the popularity of soymilk for those who maintain orthodox Jewish dietary practices is that soymilk is "Kosher" – able to be served with meat. This has led to quite a revolution in some circles, the development of many Northern European tasting soy dishes, and *tofutti*, a soy-based ice-cream which evolved from a New York Jewish café.

Tofu has appeared as a major ingredient of ice-cream in the last year. Dairy-free *tofutti* was a hit with the cholesterol-crazed public. Frankly, there are full cream ice-creams which I think are more natural and definitely better than the new tofu ice-creams. In all of the advertising I have seen for tofu ice-creams, the 'natural' and low-cholesterol, non-dairy aspects are pushed heavily, without any mention of the fact that they are as loaded with sugar as any other ice-cream, and with more chemicals than some. Sugar is called 'corn syrup' – it sounds natural, doesn't it? – but corn syrup is simply a chemical sucrose produced from cornstarch by industrial refining!

When I was first introduced to tofu in the early 1970s, the Japanese teachers warned that tofu should not, as a rule, be eaten cold in sweet-frozen desserts. Their reason was that it was particularly bad for men as it shrunk their testicles and reduced fertility! To this end, they recommended using eggs with it if chilled as a dessert, to help balance this undesirable effect. Food for thought! The effect of tofu as a food is well understood in the East, whereas we tend to use it with less discretion. When food is taken out of its cultural context, the item concerned can be altered out of recognition. In the 1960s, a middle Easterner was horrified by me putting honey with yoghurt. 'It is not a dessert!' Look what's happened to yoghurt now. One is hard pressed to find some which isn't loaded with sugar, and that hard-edge sourness has become a memory. I'm sure it wasn't the chocmint flavour which gave Bulgarian centenarians their longevity!

Soymilk virtually fits in where cow's milk does. I use it for sauces, dips, custards, cakes, bread, pastry, jellies, ice-creams, etc. It makes tremendous milkshakes with other natural ingredients such as carob, pure chocolate, honey, maple syrup, vanilla, malt. It is also a worthwhile beverage on its own and, as mentioned, very easy to make at home.

Soy Yoghurt

It is a surprise to some that dairy-style yoghurt can be made from soymilk. The process is exactly the same as for making yoghurt with cow's milk, and the product is similar to yoghurt made from skim milk, yet it is even lower in fat. The best culture I have tasted in soy yoghurt was not the familiar culture, but one known as *villia*. The taste of soy yoghurt is not quite what you are familiar with in regular yoghurt, but it is acceptable.

Dried Bean Curd or Yuba

Dried bean curd 'skin', called *yuba* in Japan, is another product from the mother soymilk. If thick soymilk is allowed to simmer uncovered at a very low temperature, a skin forms on top. By inserting a chopstick under the skin, it can be lifted out and eaten. Fresh yuba is the delicacy among soyfoods. It comes from the surface of the milk, as does cream from cow's milk, and the comparisons are really similar. Fresh yuba is creamy, sweet and delicious. Yuba is very high in protein, again without saturated fats or cholesterol.

Apart from being eaten fresh, our flap of skin flailing on the chopstick can also be dried and stored for at least six months. It is widely available in Chinese food stores as sheets, sticks or occasionally as noodles. Ask for dried bean curd sticks or sheets. The sheets are moistened or soaked and used as wrappers or containers for stuffings which are usually deep fried in Chinese cooking. Alternatively, yuba is simmered in sauces or shredded. The sticks are mainly for shredding to add a 'bone' to vegetable stir fries.

Yuba forms a fascinating part of Buddhist vegetarian cookery, wherein it takes on the guise of smoked meats, chicken slices, Buddha's 'ham', and many other meaty-textured treats.

Yuba is easy to prepare at home in limited quantities, and is well worth the effort as it can't be purchased fresh anywhere in Australia. Chinese dried yuba is, to my knowledge, dried naturally without chemicals and is a worthwhile addition to the repertoire of any cooks interested in expanding their horizons.

Soy Sauce

I used soy sauce for years without ever being really aware that it was made from soybeans. On some level, the 'soy' was synonymous with 'Chinese' — it was 'Chinese sauce'. Perhaps this is because there are no soybeans actually floating in it, and on another level, I suppose I couldn't conceive of how it was made. How did this thin, black liquid come from soybeans? It really is a revelation to discover the process, and what a masterpiece of food technology it is.

Soy sauce evolved as a development of the basic process of fermenting soybeans which originated in China in the first millennium B.C. Originally, the beans were fermented as a thick mash and gradually more and more water was added until the paste became a liquid. The process was transmitted to Japan about 700 A.D. and from there, a unique product developed which has become widely known in the West.

Of course, in the 1930s, technology entered the arena and soy sauce production was overtaken by the quick buck. A chemically-aged product was more profitable because it took a lot less than the 2½-3 years spent fermenting natural soy sauce. The results of this are the commercial soy sauces available today, which may contain sugar, caramel, preservatives, flavour enhancers, etc. These sauces bear no resemblance to the naturally fermented item which became known as *tamari* in the West. However, this is not its correct name. Tamari is indeed a soy sauce, but it is not what the Japanese use daily. Georges Ohsawa introduced the term 'tamari' to the West to distinguish natural soy sauce from the commercial concoctions. What he actually introduced, however, is what the Japanese call 'shoyu' – and what we know as soy sauce.

Shoyu is made by soaking and cooking soybeans. These are then mixed with cracked roasted wheat and inoculated with a culture. When the culture (called *koji*) has covered the wheat and soy, it is mixed with sea salt and water, placed in huge wooden barrels, and allowed to slowly ferment for 2-3 years. It is then strained to remove the soy and wheat meal, and oil is skimmed from the surface. The remains are then pasteurised and bottled.

Tamari was originally a by-product of the miso-making process. It was strained from a barrel of fermenting miso as it accumulated over the years of fermentation. Today it is made as a food in its own right, in the same way as shoyu is made, but excluding the wheat. It is made solely from soybeans, water and sea salt.

Both shoyu and tamari are soy sauces, and both, if purchased from a natural food store, are naturally made. They are really the equivalent of fine, well-matured wines in their epicurean value.

Tamari is blacker than shoyu and has what I call 'dark' flavours, which are robust and hearty. I use it for South-East Asian and Chinese-style dips and sauces; in some soups to 'extend' the miso flavour; and in baking. Shoyu seems lighter and sweeter, more in keeping with stir-fried vegetables, soup noodles, especially in combination with *mirin*, seafood dips and lighter flavours. The low-salt shoyu is very good, maintaining full flavour with half the salt content. Natural soy sauces contain their own organic preservatives and tend to preserve foods on the table. They also maintain the nutritional content of foods. For example, vitamin C, which can oxidize rapidly, remains stable for a longer period of time in the presence of shoyu.

Depending on where the soy sauce is made, it has various flavours, colours and textures. The traditional Chinese soy sauces differ from the Japanese considerably, as do the Thai and Indonesian ones. Each, of course, is dictated by climate – hotter climates producing more rapid fermentation and saltier soy sauce. There are also considerable variations among the Chinese sauces. Many of them contain unique local ingredients or are made with wheat flour or soy flour. Some contain oyster liquid (oyster sauce) and shrimps. I saw numerous earthenware pots of fermenting beans throughout rural South-East Asia, as many variations as in village wines in Europe. The Thais favour an almost camphor-flavoured soy sauce which is very thick and contains many bean particles. The Burmese use a similar sauce called *poneyegyi* which is made from the Burma bean (phasolus lunatus). In Indonesia, the soy sauce, or *kechap*, is similar to the Chinese, except for the flavoursome *kechap manis* (sweet soy), sweetened with natural palm sugar (gula jawa). This latter has a particular affinity for tempeh. In all of these soy sauce nations, one can also buy Chinese and Japanese soy sauces. These are both used along with the native produce in various combinations for many dishes.

The best natural soy sauces for everyday use are the Japanese shoyu and tamari. These are traditionally manufactured with no short-cuts, and contain the distilled essence of a thousand years of craftsmanship. They can be used in Chinese and South-East Asian cooking to great effect, and are a worthy food to use daily.

Miso

Miso is in many ways the Asian equivalent of cheese in that the fermentation processes are similar and the aims are equivalent – to render the food preserved and in fact more digestible. Miso is a Japanese term, but the process of fermenting soy beans to make a soy bean paste was learned from the Chinese over a millennium ago. It was further developed by the Japanese to suit their own tastes in much the same way as with cheese, where the Italians developed provolone and parmigiana, the French camembert and roquefort, and the English cheddar and stilton. This commonality between the animal cow and the vegetable 'cow' typifies the soy products of Asia – regional variants on the same theme which lend considerable local colour to cooking.

Miso is used extensively in Japanese cooking, and is a hallmark flavour which appears and re-appears in numerous guises. It has become very popular as a soup base, and it's not surprising, as I think properly made miso soup, using only naturally fermented miso, is one of the most calming yet invigorating foods I have every enjoyed. Many Japanese still start their day with miso soup (see recipe, p.69.)

I also use miso to make gravies, sauces, in stir frying, as a marinade, as a pickling medium, as a stock, in stews, salad dressings and a bread spread.

Japanese miso is noticeably different to its progenitor, Chinese *chiang*, which hasn't really entered Western cooking to the same extent. Miso was popularised by the natural food market. Chiang is rarely (more so in the south) used by the Chinese in soup, unlike miso, and it is this quality of each which really distinguishes them. You may know chiang from the famous sauce used for Peking duck – *hoisin sauce*, which is made from soy bean paste (chiang), sugar, vinegar, garlic and chilli. Surprisingly, hoisin made from Japanese miso is very good, in fact, better than any hoisin I've had – but I haven't visited China yet and tasted the local version.

Dark miso also substitutes for the Indonesian soy paste *taucho* (tao tjo) which is hard to get. The Burmese use a type of miso which I bought in the Shan State, but lost in Australian Customs! The Shan Burmese dry and pound it to make sauces. This variety is called *peboke*.

The regional cooking of Japan is steeped in the type of miso each locality produces. The lighter yellow misos are much sweeter than their dark counterparts. Polished rice is a major ingredient in their production and the fermentation period is shorter in the hot south. Rice miso is known as *kome* or *shiro* miso and ranges from a creamy yellow colour to russet. The rice:soy bean ratio can differ from 2:1 to 0.8:1, illustrating enormous variations in the range.

Darker misos are made from a combination of soy beans and barley, soy and brown rice, and simply soy, with no cereal. Barley combinations produce *mugi* miso, deep red-brown, with a hearty, semi-sweet flavour. Brown rice or *genmai* miso is my favourite soup miso — yellow-brown, savoury sweet and extremely delicious. I keep a keg of this which goes darker by the month and changes from a light flavour to a very strong flavour as its colour changes from light brown to almost black.

Hatcho miso is black-brown and contains no cereal, only soy. It has a vastly different flavour to any other miso and works well in creating Chinese-style dips and sauces. Hatcho miso comes from a region renowned for Samurai and the locals attribute the strength and valour of their mediaeval warriors to the quality of their miso. There are other varieties of miso available which contain various cereals such as buckwheat instead of rice *(soba* miso) or pickled vegetables and barley malt (*natto* miso). These are the varieties generally available in Australia. To the Japanese, my description is ridiculous as there are over 200 varieties of miso with many blurred boundaries. The Japanese also mix misos to some extent, eliciting new flavours again!

Miso is not hard to make, but one needs to be well prepared and very clear on procedures. I have made excellent rice miso on my first go, and I've sampled good barley miso made in Australia. In the USA, very high-quality miso is being produced using some innovative ingredients. Chickpeaso is made from chick peas and corn. Experimenters in India have made it from lentils and peanuts! This is indeed an interesting culinary event for the West, to have adopted an Asian food and developed it. The American misos are all natural and quite a tribute to the people involved.

Miso must have something going for it! It is especially the naturally fermented misos that are finding favour amongst those tired of a diet with too much meat, fat and sugar. These natural misos, being high in protein and having a noticeably invigorating quality, are a perfect addition to one's diet if it is becoming lighter on meat.

Whenever miso is used as soup or on grilled tofu or fish, it lends a robust quality which provides the same sort of satisfaction as animal products. I have used miso very successfully as a soup stock where meat stocks may be the conventional base. Hatcho miso lends extraordinary body to minestrone, for example, where traditionally a meat stock is employed.

It is beyond the scope of this book to provide methodology for miso making, but a short description of the process involved will give an insight into its simplicity, yet technical mastery. Soy beans are soaked and steamed; the accompanying cereal – whether it be rice, barley or buckwheat – is also cooked. These are then mixed and a yeast-like culture is added. The culture takes about 3-4 days to grow throughout the mix. When it has sufficiently spread, sea salt is added and the mass is placed in huge cedar vats. The lid is placed on the vats and stones are piled on to the lid to create immense pressure. Depending on the season, the proportions and the style of miso, the mix is then fermented untouched from a few weeks to three years. A dark amber liquid, which separates from the fermenting miso, is sometimes drawn off through spigots. This is authentic tamari and is in itself an extraordinary by-product of miso making.

A great contrast is evident if one compares the above process which requires considerable skill and craftsmanship, to modern miso-making methods which can be completed in one week. Often de-fatted beans are used, acids and chemicals are added to break down the soy beans rapidly, and flavours and colour components are added, rather than being allowed to develop over the years from complex fermentation. Such chemical misos are generally available from Japanese food stores, and I recommend you avoid them. Search out the authentic items from 'health' food or natural food stores. Miso, as a naturally produced, traditionally fermented whole food, has considerable virtues to offer us as a culinary ingredient, with a tremendous taste in its own right.

The packaging of miso necessitates that the miso be pasturized. This is to prevent the packages swelling up and exploding in transit. Unfortunately, the pasteurization destroys a lot of the good work done in traditionally preparing it, and pasteurized miso is definitely less nutritious. Some importers purchase kegs of miso and re-package it in plastic tubs. Sometimes this is labelled 'unpasteurized'. It is the best way to purchase miso.

Tempeh

In Indonesia, cooked soy beans are inoculated with a starter, wrapped in banana leaves and left to ferment in the humidity. When each little parcel is unwrapped, the beans are bound together by a 'furry' mushroom-like culture which has grown through them.

It is curious that almost all of humankind's attempts to make the soy bean digestible have involved fermentation. Some processes borrowed from other cultures – as the Japanese refined their idea of miso and shoyu, quite distinct from their Chinese parents – the Thais and Indonesians make their own varieties as well. However, no culture has produced an equivalent for *tempeh* – it is a unique product. In fact, the most similar product I have encountered is something very far from Asia and that is the soft, ripened cheeses of France. Tempeh is a cake of cooked soy beans bound together by a white mycelium (mould), much in the same way as camembert or brie have soft, white mantles. The spore culture mainly responsible for this is rhizopus oligosporus. This mould actually penetrates the beans to an extent, and renders them more digestible. As a by-product, the mould itself and others which play a less important role in the process, are particularly nutritious, especially because vitamin B_{12} is synthesized. Apart from having B_{12}, the profile of cultures which make up the mycelium contain many other nutrients as well. But tempeh is particularly prized because of its complete protein.

All this is biochemical and doesn't mean anything until the tempeh goes into one's mouth – and it's delicious. The flavour has often been compared to mushroom, and this is true, but a lot depends on the method of preparation. Tempeh is almost universally fried in Indonesia, and usually deep fried. Here it's at its best I feel, particularly with the delicious marinades used before the frying – tempeh goreng (tempeh fried) is served throughout Indonesia and varies delightfully from one area to the next.

I'm not one to go along with frying paranoia. If you consider frying to be unhealthful, then your choices with tempeh are few and, in a culinary sense, very limited. If it doesn't bother you to eat fried food, then you are about to experience some flavour treats. I rate fried tempeh as just about the closest thing to meat the vegetable world has offered us. Its texture, its flavour and is high protein value make it a very satisfying food.

Tempeh produced outside Indonesia is slightly different to the Indonesian product. This is because many flavours are created in the uncontrolled (wild, natural) fermentation which we do not taste in the monoculturing needed to produce tempeh in a cooler climate. Tempeh culture will only work inside an incubator outside the tropics, so the culturing has to be very controlled, not allowing for the growth of other organisms. Studies have shown the mono-cultured tempeh from an incubator to have less of the vitamin B_{12} of a wild Indonesian sample.

I prefer to eat tempeh in summer, as it is a hot-climate food. There is something unsettling to me about eating tempeh made in the middle of winter in an incubator – it's similar to eating out-of-season vegetables. According to Chinese medicine, fresh tropical produce has a cooling effect – commonsense, in the same way yak butter heats one up in Lhasa. To eat mangoes and the like during a winter is inviting this cooling effect and consequent metabolic imbalance

— we call it 'a cold'. The condiments which most enhance tempeh — chilli, coconut milk, cucumber, raw and blanched vegetables — certainly sound cooling, chilli paradoxically so as it seems to engulf one with fire! Hence, generally speaking, tempeh lends itself best to summer cooking.

Tempeh is available prepacked in plastic from 'health' and natural food stores and a growing number of oriental food shops. It is either frozen or simply kept refrigerated. It will keep refrigerated for 2 weeks. I actually prefer it a little older. Sometimes I leave the opened package unrefrigerated for a day. If a dark shadow of sporulating culture appears on the surface, simply wipe or rinse it off. If the shadow hasn't grown to a fuzz, I use it as is. The wrapper of 'Nectar' tempeh, actually states: 'It may develop darker areas indicating maturity. This colour is in no way harmful and in fact increases nutrition and flavour.' Well said. Vitamin B_{12} probably increases due to this, and I agree, the flavour is much better.

On a culinary level, tempeh is exciting, not only if you eat little meat or dairy foods, but even if you are a jaded cook looking for inspiration. Besides deep or shallow frying, which I think really suits tempeh, it can be pureed into dips, steamed, or even incorporated into bread. But in many ways it is the Indonesians' cooking of tempeh and its inclusion amongst delicious and exotic ingredients which really elevates it to culinary greatness. Tempeh fits surprisingly into Western dishes or one can create East-West fusion dishes from it. As such it deserves a place in our growing planetary cuisine.

NUTRITION AND SOYFOODS

The only soy products we had seen in the West until the 1970s, apart from those available in Asian food stores, were the textured vegetable proteins (TVPs) and other canned and packaged 'vege' foods. It seemed we had been sentenced to a bleak future munching on fibre pills and TVPs while tripping around in metallic jump-suits. The best our highly sophisticated industrial methods could wring from the soybean were such unappetising pseudo-foods. One of the miracles of soyfoods is that they can all be made in the home kitchen – they require no sophisticated processes which inevitably de-nature foods. And soyfoods are actually 'food', not substitutes like TVPs.

With all prepared foods, the degree and manner of processing is highly significant. For example, let's take two bottles of soy sauce – samples A and B – with the same ingredients: soybeans, wheat, salt, water. What the label doesn't say is that sample A used de-fatted soybean meal which was broken down by industrial acids. The wheat was actually white flour which may have been caramelised or turned into sugars, the salt was raw sodium chloride, and who knows where the water came from and what it contained. Sample B may have been made with whole soybeans and wheat, with sea salt and spring or well water. Labels don't say how the food was made, and many of us would prefer not to know. How it was made is as important as what's in it. The degree of processing is what distinguishes a natural product from an industrial one.

Traditional food processing actually enhances the food value of soybeans. Over many centuries, craftspeople have learned how to carefully manage fermentation so as to produce the most palatable foods. In the fermentation of soybeans to make miso and shoyu, myriads of micro-organisms, ferments, enzymes and free amino acids are produced.

Research on shoyu has been considerable in Japan, and the results are fascinating. The fermentation is very similar to human digestive processes, releasing nutrients which are virtually 'pre-digested' and easily assimilable. The enzymes which are particularly active in the fermentation become strong digestive agents when they enter our digestive tract and they stimulate the secretion of digestive fluids. In other words, tamari and shoyu, besides being very assimilable foods themselves, assist in the digestion of foods they are eaten with. The same research also indicates that chemical soy sauces do not have the nutrients or enzymes which the traditional product has, and furthermore, tend to inhibit secretions in the digestive tract.

Similarly in miso, lactic acid bacteria (lactobacillus) aid digestion and the absorption of nutrients. This is because nutrients are broken down or unlocked from complex structures and are then easily absorbed. This is particularly important for proteins because they are utilised more efficiently than by eating a steak, which requires considerable internal effort. A proportion of such 'raw' protein will be excreted or turned into toxic substances which deposit in joints or maraude the bloodstream. Thus miso is a natural miracle in that one gets far more nutrition from it than from eating an equivalent amount of soybeans. As with shoyu, commercial miso has the opposite effect to its naturally fermented counterpart – none of the benefits of the lengthy fermentation are available to you, and it doesn't taste good!

Tempeh fermentation is equally significant in that the beans are rendered more digestible and new nutrients are created. One aspect of tempeh research which parallels work on naturally fermented sourdough bread is that the fermentation produces the enzyme phytase, which hydrolizes or breaks down phytates. These are substances which tie up minerals such as calcium and zinc. Thus the fermentation has made these nutrients available to us, where they weren't in the 'raw' food.

An important aspect of soy fermentation is the synthesis of vitamin B_{12}. Shoyu and miso contain it in small, varying amounts. Tempeh contains much larger quantities and some samples had as much as 400 per cent of recommended daily allowance (RDA). Tempeh also contains anti-bacterial agents which act as antibiotics to organisms such as streptococcus aureus (see Shurtleff), which is notorious for causing food poisoning. Similarly, some cheeses produce antibiotics; our penicillin was discovered in this way.

One of our looming problems as we embrace more refined foods is that we miss out on the micronutrients and other nutritive factors such as trace elements which were unlocked by our traditional ferments. Pickled vegetables, ales and stouts, cheeses, vinegar, wines and bread were all once made with natural fermentation. Now, very few – if any – are made in the time-honoured fashion, and it is important that they be revived as aspects of both nutrition and culture.

Soybeans certainly earn their title – 'meat of the fields'. Among vegetable foods, they are high in protein and score very well in comparison to animal foods. 'Protein' is an empty term with little meaning as a dry statistic. What the term usually refers to is the percentage by weight of protein that a food contains. Soybeans rate very well on this score. Dried yuba is 52 per cent protein by weight, with dried soybeans at 35 per cent, cheeses 30 per cent, fish 22 per cent, beef 20 per cent, tempeh 19.5 per cent, tofu 8 per cent, and cow's milk 3 per cent. But remember, this protein is basically locked up in the soybean; processing releases it. This leads to a more significant statistic and that is net protein utilisation (NPU) which means, in short, how well that protein can be used by the body. Eggs rate highly with an NPU of 94 per cent, fish 80 per cent, beef 67 per cent and tofu 65 per cent. Brown rice rates 70 per cent, although its percentage by weight is only 6 per cent! If we combine the quantity of protein, i.e. percentage by weight, with the quality, i.e. the NPU, a whole new picture emerges. It is the quality of protein which is a more significant figure for us, and, as you can see, tofu, with only 8 per cent by weight of protein yet with 65 per cent availability, is a more efficient food than meat, which has 20 per cent by weight and 67 per cent availability. A 250 g block of tofu provides about 12 g of usable protein. This is approximately 30 per cent of the RDA and is equivalent to roughly 100 g of steak, 2 eggs, 60 g hard cheese or 2 cups of cow's milk.

Protein is composed of amino acids: there must be eight particular amino acids present in a food for it to be complete protein, i.e., needing no supplementation. These are the essential amino acids. The soybean is the only legume which has complete protein, as does meat, for example. Cereal grains such as rice or wheat have limiting amino acids. That is, they are deficient in complete protein because there are certain amino acids they don't have, which limits them as a protein. But if we combine protein from a legume (soy) with a cereal (rice), not only do the proteins match each other's deficiency and become complete, but more new protein is created. The limiting aminos in legumes are usually methionine and cystine; in cereals it is lysine. Soyfoods are high in lysine, but limited in methionine and cystine. Thus they are the perfect complement, increasing protein by up to 42 per cent if served together in an appropriate ratio, which is usually 4 parts cereal to 1 part legume.

The diets of all traditional peoples are mirrored in this combination. The legume cereal complement has sustained humanity for thousands of years. For example, in India, the combination of wheat or rice and dahl (legumes), quinoa corn and beans in the Americas, millet and beans in Africa, rice and soy (tofu, miso) in East Asia and rice tempeh/tofu in South-East Asia, and bread and 'pease' in Europe. All of these cultures then supplemented these combinations with small amounts of meat or dairy. This is exactly opposite to the way we have come to eat today in the Western world.

When protein sources are compared, soyfoods begin to rate very highly. Now consider their other components. Animal and dairy products contain quite high levels of saturated fats and cholesterol, whereas soyfoods contain no cholesterol and very little saturated fat. What soyfoods do contain is linoleic acids and lecithin, which break down cholesterol and fat deposits in the organs and blood. This is significant because those of us who need to get rid of these deposits will derive double benefits from soyfoods.

Being high in the food chain, animal and dairy foods are relatively high in chemical toxins – pesticides, agricultural chemicals, fertiliser residues, antibiotic and medicament residues. Milk is a classic accumulator of radio-active elements, for example, and the press has increasing reports of alarming levels of residues found particularly in meat. Plant foods, being very low in the food chain, contain very few of these toxins, especially if one uses organic produce. This is another example of 'how it is made' being more important than 'what's in it', and underlines the growing benefits of soyfoods as a protein source.

More than half the world's population has difficulty digesting dairy products, particularly non-northern Europeans. But even 8-10 per cent of Europeans, especially infants, cannot digest cow's milk. These people are lactose intolerant, which means they lack the enzyme lactase, necessary to break down lactose or milk sugar. It seems that lactase disappears from the digestive tract on weaning. The East and South-East Asians are unique in that they have never used dairy products as food, although historically, they had plenty of opportunity to do so. Their milk comes from the vegetable cow — soymilk — and there are no problems digesting it.

East Asian traditional medicine has many fascinating perspectives on drinking cow's milk. It is generally believed to extend one's childhood physiology; this is mirrored in the (natural) disappearance of lactase after weaning. Because physiology and psychology are inextricably linked in this medical system, it is regarded as inappropriate or dysfunctional to continue drinking any form of milk after weaning.

The majority of peoples who use milk traditionally, always process it (as cheese, yoghurt, butter), and our modern milk drinking really has no historical equivalent. Soymilk, having all the advantages of being low fat, cholesterol free, easily made, more efficient from a land-use perspective, and palatable, is creating justifiable paranoia among the dairy industry.

For our ever-increasing planetary population and dwindling resources, soyfoods have yet another role to play. From one acre of land it is possible to produce 20 times more usable protein per acre by growing soybeans than by animal husbandry! The recent trend in feedlot animals or grain-fed beef means that our soybean crop is being wasted in the most extravagant fashion. It takes about 11 kilograms of soy protein to produce about 700 g of beef protein if the soy is fed to animals à la feedlot. One doesn't have to be a genius to extrapolate that if the US soy crop (their *biggest* agricultural crop) and ours were used directly as human food, planetary nutrition would be sufficient. This also means that the wasteful clearing of land for grazing and the miles of seemingly empty and useless countryside could end, allowing reafforestation and a bit more clean air for us all! 'We diet to curb obesity while they starve . . .'

I include all the above information to clarify, if possible, the place of soyfoods. I know many people still consider animal protein to be superior to vegetable protein, and this is typical of the sort of misinformation we use to form our picture of the world. I am not a nutritionist, and nutrition doesn't fascinate me at all. Modern Western nutrition is a fledgling study which does not integrate its findings significantly and generally misleads us all. What's more, few medical

practitioners are aware of nutrition, and the actual results of nutritional stu...
don't permeate society because of this. Of course, vested interests in 'junk' foo...
have no small part to play here in funding the right answers.

I continually emphasise the theme that we are re-inventing the wheel as far as diet and nutrition are concerned. Those in the forefront of nutritional research validate all the principles of diet which have been well known in the Orient for millennia, i.e. basic foods should be cereals, legumes and vegetables, complemented by small amounts of meat and/or dairy. This is not only a nutritional consideration, but has considerable ecological and aesthetic ramifications.

INGREDIENTS AND TECHNIQUES

There are many seemingly exotic ingredients in this book, but they can be obtained in any city if you really look. I have outlined exactly what they are in the Glossary. The ingredients I use are all 'natural' — isn't it odd they have to be termed so?

The term 'sea salt' refers to salt which comes from evaporated sea water. It is not treated with chemicals or free-flowing agents (I have found eight such ingredients in some salt), and is not pure sodium chloride. Sea salt contains a tiny proportion of micro-nutrients in the form of minerals and trace elements, and is a different product. I recommend you use it exclusively. All salt in the recipes is optional, so you may exclude it.

The oils I use are:

1. *Chinese Peanut Oil* — this has an appetising aroma of peanuts with good colour and not too thin texture. Not all Chinese peanut oil is like this; the one I use comes in a 3-litre tin with 2 red lanterns on the front. It is excellent for deep frying. To keep deep-frying oil, strain well, put in a glass jar with 1 or 2 umeboshi plums. You won't believe how well it keeps.

2. *Light Sesame Oil* — cold pressed from natural food and 'health' food stores. Good for sautéing.

3. *Extra Virgin Italian Olive Oil* — for salads and 'European' flavours.

4. *Cold Pressed Safflower Oil* — for its mild flavour.

5. *Dark Sesame Oil* — for flavouring only, at the end of stir frying.

'Honey' refers to pure, cold-extracted honey, not the supermarket product; the two are very different. 'Prepared mustard' is any good quality mustard which is ground with vinegar and/or white wine and without unnecessary additives.

Free-range eggs are well worth using, even if they aren't quite 'free'. They're still better than supermarket eggs. I first became aware of this when I noticed how beautiful the eggs were in rural Southern Thailand — deep, rich yolks and almost opalescent whites. Free-range eggs generally have firmer yolks and better white texture.

I use freshly ground spices whenever possible. The difference in flavour to the store-bought powder will be obvious once you start grinding your own. Invest in a stone mortar and pestle which are widely available in Chinese stores.

Agar flakes or bars or strips are better than agar powder, which is manufactured chemically. The flakes are very convenient, but the strips are usually much cheaper.

There are some techniques that need explaining:

Tamarind water: Simply make a paste with water from the block and separate the seeds. 1 cup water to 1 heaped T of tamarind.

Coconut cream: This must be made from sound coconuts. They shouldn't be leaky or dry or slightly fermented smelling, but hard, with a good water content. Break up the coconut and separate shell. Peel off brown skin and purée flesh in a blender or food processor with water – about 3 cups per coconut. Strain and squeeze the pulp well. If you let it stand, a cream will form on the top. This can be scooped off and used for rich sauces, while the remaining water is excellent in soups, stews or light sauces – or to drink in smoothies. I generally use the lot, mixed together. When cooking, make sure to stir while heating coconut milk or it may curdle.

Kechap Manis can be made by combining 1 cup palm sugar with 1 cup water. Bring to a boil and simmer, stirring until all sugar dissolves. Turn off heat, add 1 cup tamari, 1 star aniseed, ½ t Laos, ½ t coriander seeds coarsely crushed, 4 T rice malt or maltose. Stir until well mixed. Cool, bottle and keep in a cool place at least one week before using.

Agé is generally oily when you buy it. Put in a bowl and pour boiling water over it and squeeze with a wooden spoon. Lots of oil comes out. Drain and use.

Tofu often needs pressing before use. Place a block on a dry cloth on a chopping board, wrap the cloth around it and put something mildly heavy, like a small chopping board, on top for 10-15 mins, then pat dry. A 'block' of tofu means 450-500 g.

Lemongrass can often be bought in stalks which still contain some root stock. Take off the coarse outside leaves, chop the lot into 1 cm pieces. Put in a mortar and pound it, just to break it open, or use a rolling pin on a chopping board.

Laos often comes in whole, dried root pieces. These can be grated on the nutmeg grater and give the best flavoured Laos. Same for *tumeric*.

If you don't have a blender, push *tofu* through a fine sieve with a wooden spoon and then whisk it smooth.

Tempeh can be blanched in boiling water with a little shoyu. This is a much more acceptable flavour than plain water blanching.

Miso, contrary to popular opinion, does not need refrigerating and, in fact, keeps much better out of the refrigerator. If it grows a white mould, simply stir it back in. *White miso*, however, may go 'off' in a very hot weather.

To cook *short-grain brown rice* (I always use biodynamic rice – it tastes best):

2 cups rice   pinch salt
4 cups water

Wash rice well by rubbing between your hands under water. Drain well, rinse, add to cold water with salt. Put the lid on, bring to a boil and then simmer slowly for at least 45 mins without taking the lid off at all. It doesn't matter if the bottom is scorched (not burnt!), as this is a sign of well-cooked rice and the scorched part is delicious.

Long-grain brown rice:

2 cups rice   pinch salt
3½ cups water

Measure out water and bring to the boil. Wash rice. Drain well. Add to boiling water, cover and allow to re-boil. Simmer 30 mins.

Thai jasmine rice or basmati rice:

2 cups rice   pinch sea salt
3 cups water

Wash this rice very well. Rub it until the water rinses clear. Put rice and water in pot, cover, bring to the boil. Simmer for 15-20 mins without removing the lid. Or, with the same measurements quoted here, use the technique for long-grain brown rice.

Black rice:
Soak overnight. Rinse and discard water. Put in saucepan, just cover and bring to the boil. Simmer covered for 30 mins. Add more water for 'stickier' rice. The technique for white, glutinous rice is the same.

Millet:

As for short-grain brown rice. Up to an extra cup of water can be used for a fluffier grain.

Barley:

As with short-grain brown rice.

To cook hijiki:

Wash then soak in cold water for 30 mins. Scoop out hijiki with hands, leave the water behind, and place in saucepan. Carefully pour in just enough soaking water to barely cover hijiki. Be careful not to drain this soaking water as it is sandy. Bring to the boil. Add shoyu – smell the cooking hijiki – add shoyu little by little until the shoyu aroma just dominates the hijiki. Re-boil and simmer rapidly until almost all the water is evaporated. Strain. Bigger, branchier hijiki is best.

To cook noodles:

Refer to section on *soup noodles.*

Kombu stock (1):

Soak 2 sticks kombu seaweed in 1 litre cold water for $\frac{1}{2}$ hour. Slowly bring to the boil. Turn down heat just before boiling and simmer slowly for 10 mins. Remove kombu and use stock. Kombu can then be soused with vinegar and sliced for salads or as a garnish.

Kombu stock (2):

Repeat above procedure but add 2 small or 1 large shiitake mushrooms.

Kombu stock (3):

Repeat original procedure, but when water almost reaches a boil, add 2 T bonito flakes and 1 cup cold water. Simmer for 15 mins. Don't boil or it will be bitter. Strain and discard bonito.

Instant dashi powder can be bought in 'health' food and natural food stores. One type has kombu, shiitake mushrooms and shoyu. The other contains bonito. Substitute $\frac{1}{2}$ sachet in the above recipe, i.e. 1 litre water, $\frac{1}{2}$ sachet instant dashi.

Occasionally, *black sugar* is mentioned. This is simply pure sugar cane juice which has been reduced by boiling. Only the water is removed. It contains its full complement of minerals, fibre and colour, unlike white sugar. It is available from Asian food stores in a dish or cup shape, usually imported from the Philippines.

Curry powders can vary tremendously. The ones most suitable for use with these recipes come from Singapore, Malaysia or Sri Lanka. The Malaysian ones vary from brown to yellow-brown. The Sri Lankan ones are a dark brown powder, containing pieces of herbs; they are roasted. It is worth poking around and trying different ones. I have discovered some beauties this way – often never to be found again!

Prepared chilli refers to a simple chilli paste, usually containing only chillies, water and salt.

For making sauces such as *nam prik* (p.64), onions, chilli and garlic are usually seared over coals, cleaned and pounded together with other ingredients. This exemplifies the technique in South-East Asia, and often accounts for an elusive flavour you may not have been able to catch.

Utensils

To undertake deep frying you really need appropriate tools: a *deep-frying pot, wok or saucepan with a strainer*, preferably the type that simply clip on to the side. *Paper towels* and a secondary drainer such as a *cake cooking rack* are useful. Stand it on a *cake or biscuit tray*. A good pair of *tongs and a wire mesh scoop*.

To cook cereal grains requires a *heavy, non-aluminium pot*. The heavy French or German enamel ware is perfect for whole cereal grains. Stainless steel is all right as long as it has a good, thick base.

Stir frying is best done in a wok with the appropriate *wok shovel*. A wok with a single handle is a useful instrument. The wok needs a *lid*. Some domestic ranges especially electric ones, don't allow for stir frying because not enough heat can be generated from them. *A gas burner* is a good investment, and it can be taken anywhere.

Good *saucepans* are valuable assets. Avoid aluminium and non-stick. Stainless steel is good for boiling, while enamel is better for sauces and cremes.

The ingredients mentioned in this book are generally available in Asian food stores, 'health' and natural food stores.

Soyfoods may be new to many of you and so I have concocted a few menus to illustrate how they can be included in your meals. These are fairly concentrated soymeals and I'm not suggesting any sort of soy regime – but they indicate where soyfoods can fit in.

WINTER
Breakfast
Miso Soup
Bread or toast with spread and white miso
Cereal or porridge with soymilk

Lunch
Tofu or tempeh burgers with lettuce, cooked beetroot, carrot, sprouts, tofu mayo and pickled onion or gherkin

Dinner
Lentil-miso soup
Brown rice
Stir-fried greens, baked vegetables and tofu (with shoyu)
Salad
Soy pipeline custard

or

Breakfast
Miso congee
Bread or toast with spread and white miso
Glass of soymilk

Lunch
Buckwheat noodle and tofu soup

Dinner
Tofu and tempeh kebabs with teriyaki
Millet, stir-fried vegetables
Blanched salad, soy semolina

SUMMER
Breakfast
Cereal, granola or muesli with soymilk

Lunch
Noodle soup with tofu (and chillies)
Soymilkshake

Dinner
White miso soup
Tempeh goreng with a char
Thai jasmine rice salad
'Blueberry Sunset'

or

Breakfast
White miso soup
cereal, granola or muesli with soymilk

Lunch
Grilled tofu and salad sandwich

Dinner
Tempeh mee-goreng
Salad
Tofu mousse with strawberries

How to make Tofu and Soymilk

Making tofu is no more difficult than making bread. Most people simply assume it is complex. At first it is time consuming and messy, but after four or five attempts you can reduce this time to 1 hour. The proper equipment is absolutely essential or a mess and confusion can result.

Basically, you need a large pot, 2 big bowls, a vitamiser, a cloth sack (flour bag), cheesecloth and a settling container.

Use good quality, organic soy beans if possible. Soak them in plenty of water overnight. After soaking, strain off the liquid and wash the beans. The following measurements are for 1 cup soaked soy beans. Multiply as necessary.

Bring a kettle of water to the boil. Place 5 cups water in a large pot over high heat. Meanwhile, purée (vitamise) the beans with 2½ cups boiling water until well ground and milky. The blended beans are called raw gô purée. Add this purée to the 5 cups of hot water. Stir continuously over medium heat until the mixture is well cooked and foam rises in the pot. The product at this stage is called gô, and can be used in other receipes. Pour the purée into the moistened cloth sack, set in a bowl. Strain the liquid and press back very firmly to extract all the soymilk.

Place the sack in 1 or 2 cups heated water in a saucepan and squeeze or press again to ensure that all the milk is extracted. The grounds left in the sack are called okara. Remove and keep for other cooking. Place all the soymilk back into the cooking pot and bring to a boil over high heat. Stir often so the milk does not burn on the bottom of the pot. When the soymilk reaches boiling point, turn down heat and simmer slowly for 5 minutes. This can now be used as soymilk.

Now curdle the soymilk to produce tofu. This is done by using nigari or clean seawater. These produce the best quality tofu. Use 1 cup seawater for each cup of soaked soya beans used, or 1½ teaspoons finely ground nigari dissolved in 1 cup of water. Remove soymilk from heat. Stir it vigorously a few times and carefully pour in ⅔ of the solidifier. Stir gently to ensure even distribution. Pour the remaining ⅓ cup of solidifier over the surface of the milk. Cover and wait for a few minutes – or gleefully watch the delicate curds forming.

Give the curdling mixture a very gentle stir. By now you should have a lattice of curds floating in clear liquid. Put a large strainer into the pot to enable you to gently ladle out as much of the clear liquid (whey) as possible – you should be careful not to break up the curds.

Next, carefully pour the curds and remaining liquid into the settling container which is lined with moistened cheesecloth. The curds will remain on the cloth and the liquid will drain off. Fold the cloth over the top of the tofu and place the top of the settling container on this, then a weight, and leave for 2-3 hours. Remove weight, unwrap tofu under cold water and there it is.

Yuba

You will need a large enamel roasting pan for this, and a similarly shaped but slightly smaller pan. This works as a double boiler. Pour boiling water in water bath and heat over medium heat – it should be hot but not boiling. Pour 40 mm soymilk into smaller pan and place in water bath. The soymilk will heat up and steam. Very soon a skin forms on the milk. Use a knife to make sure the skin hasn't adhered to the pan, slip an oiled chopstick under the skin and lift out. Place on a moistened tea towel to form a sheet or put into a bowl. Repeat the procedure until soymilk is used up. The last piece of yuba is the best. It's best to eat fresh as it comes.

Silken Tofu (tahu fah)

Follow tofu-making procedure. But use an extra ½ cup soaked soybeans for a thicker milk. When curds form, leave them in the whey and allow to settle completely (30 minutes). Ladle out whey, being careful not to damage the curds. The tahu fah can then be scooped out into bowls.

Alternatively, heat 3½ cups soymilk. Dissolve ½ teaspoon gypsum powder (calcium sulphate) in 2 tablespoons of water. Pour them simultaneously into a large bowl or casserole – deeper than wider – stir gently for a few seconds. Stand spoon up in turbulence and wait until it subsides. Lift out spoon and proceed as above.

Making Agé Tofu

Japanese style:
Cut tofu into 10 mm slices – larger if you want to stuff them, 100 mm x 50 mm or 150 mm x 50 mm for giants. Place on a dry cloth, cover with another cloth and pat them gently. Change cloths until tofu is quite dry. For deep frying, heat enough good-quality oil, preferably Chinese peanut oil, until ready – drop in a tiny piece of tofu; it should sink to the bottom and return to the top immediately. Slip 1 or 2 pieces of tofu in and deep fry until golden. When cool, roll gently with a rolling pin. This makes it easier to open. Cut the top off or slice in half, poke a chopstick in and gently separate.

Chinese style:
Deep fry well-drained and pressed 3 mm cubes of tofu until golden brown. Alternatively, cut into triangles with a 6 cm base and deep fry until golden.

Breads

Moist and flavoursome breads, even croissants, can be made using tofu and soymilk. I have already dealt with this subject in the *Natural Tucker* bread book, so if you are interested, please consult it.

GLOSSARY

Agar
Seaweed gelatin. Low in calories, high in minerals, it makes a beautifully textured jelly which will set at room temperature. Comes as bars, flakes or strands. Avoid the powder as it is chemically manufactured. Agar is very cooling in hot weather and seems to promote strong nails.

Apple Concentrate
A syrup made by concentrating apple juice, often industrially.

Arrowroot
The starch from the root of a sub-tropical plant. It is used as a thickener. Kuzu is often called arrowroot, but they are totally different products.

Asafoetidia (Hing)
Keep it well wrapped! A strong, sulphurous aroma. Hing is basically a gum exudate from a tree. It is widely used in Indian cooking, particularly with *dahl*, as it is an anti-flatulent. It was also used by the ancient Romans.

Azuki Beans
Small red beans which are similar to mung beans. Used widely in Asian cooking, especially in sweets. They are a heating bean, used in Chinese medicine for weak kidneys.

Besan
See *chick pea flour*.

Blachan
A dark paste of fermented shrimp. Some people won't have it in the house because of its strong, challenging smell. Be sure to wrap it well with plastic or store in a glass jar, otherwise everything in the refrigerator will taste and smell like it. Blachan is used in very small quantities. Vegetarians may substitute twice the quantity of dark miso for blachan and ½ cup of crumbled toasted nori. It won't be the same, but it's still good.

Black Beans
Small, fermented and dried black soya beans, which are the basis of *black bean sauce*. They have a strong aroma and salty flavour, so use moderately.

Black Sugar

Raw, unprocessed, unbleached cane sugar – not 'raw' sugar – black sugar has simply had the water removed by boiling. Called *gur* in India.

Bok Choy

One of the numerous Chinese greens. It has thick white stems and dark green leaves.

Bonito Flakes

Bonito, a large, tuna-type fish, is salted and dried. Then it is sawn into blocks or pieces. These are then grated or flaked on what resembles an upside-down carpenter's plane. The resultant flakes are used as a dashi or stock ingredient.

Carob Powder

The dried and ground beans of the carob tree. Naturally sweet, carob is a chocolate substitute although when properly used, is a food in its own right. The best way to use carob is as *carob molasses*. I have only ever found one type made in the traditional fashion. Most are just carob and sugar. The traditional molasses, or syrup, is made by pulverising carob beans and steeping them in water in clay jugs. The water soaks through and is allowed to drip out. It is then tipped back through the carob. This is repeated numerous times. The dark liquid is then boiled down to concentrate it. It is a really delicious natural product which has its own unique flavour and sweetness. You may find it in a Lebanese store, or ask an importer to import it.

Chick Pea Flour (Besan)

Besan is not strictly chick pea flour, as it may have other lentil flours mixed with it, but that's how it's marketed. A fine, soft, yellow flour, highly suitable for deep frying, where it imparts a crisp-textured, reddish batter.

Chocolate Powder

Pure chocolate powder is made by roasting and grinding *cacao seeds*. It is then processed further (I don't know how) to yield pure chocolate powder. If you have never used it, you must try it as it is the deepest, most intense chocolate experience without the sugar, emulsifiers, milk, etc. Continental delicatessens or Italian food stores usually carry it.

Curry Leaves
Called karapincha, used in Southern Indian and Sri Lankan cooking to impart a mild, pungent flavour.

Daikon
Long white radish used widely in Asia and in parts of Europe. Has a huge amount of vitamin C and is prized as a digestive. Hence it accompanies deep-fried foods in Japan. Called *lo bak* in Chinese (Cantonese).

Dashi
Japanese stock. Made from kombu and/or shiitake mushrooms and bonito flakes.

Dried Shrimp
These are tiny shrimp, salted and dried. They can be powdered in a blender or mortar and pestle, or soaked or simply fried with the onions and garlic. Available in Asian stores.

Garam Masala
A basic mixture of spices used in Indian and some Malaysian cooking. Usually contains cloves, cinnamon and cardemom.

Ghee
Clarified butter. Available in tins or make it yourself by gently heating unsalted butter and continually skimming the milk solids which rise to the surface.

Hijiki
A black, stringy Japanese seaweed, particularly rich in calcium. Another strong maritime flavour which has a lot of fans. Its colour is useful in presentation, easily counterpoised with others.

Kechap Manis
Sweet soy sauce made in Indonesia, often in the traditional manner, from black soy beans, water, spices and palm sugar. Read the label carefully. An authentic flavour in Indonesian dishes and excellent with *tempeh*.

Kombu
A seaweed from the north Pacific, high in minerals, used to make stock as it is a natural flavour enhancer.

Kuzu

A high quality, naturally extracted thickener. This is washed from the huge roots of a plant growing in the cold mountainous regions of Japan. It forms a sauce of imcomparable texture. It is also a notable digestive, being soothing and nourishing.

Laos

Rhizome of a plant similar to ginger. Sometimes called *galangal*. Used extensively in Indonesia and Thailand and to varying extents through South-East Asia. Was a popular ingredient in mediaeval cooking in Europe.

Lemongrass

Also called *citronella*. Grows extensively in warm parts of New South Wales and Queensland. Imparts fragrance and flavour to stocks and is very cooling.

Lotus Root Powder

Powdered lotus root has an unusual spicy flavour which can be used to great effect in certain sweets. It is a traditional medicinal specific for congested lungs.

Maltose

A similar product to rice malt except it is made in part from wheat.

Mandarin Rind

Keep the rind from unsprayed mandarins and dry it naturally for about 6 weeks. Then chop it finely. Lime rind can be dried in the same way.

Maple Syrup

The sweet sap of the maple tree which is simply boiled down to concentrate it by removing water. Be sure to get the pure variety and nct 'maple-flavoured' syrup, which is just flavoured sugar. Pure maple syrup has a unique flavour and, except for its price, is a good sweetener.

Mirin

Cooking wine made in the traditional manner from fermented sweet or glutinous rice. Mirin flavour is a characteristic Japanese taste featured, for example, in *sukiyaki*. It has a sweetness that harmonises flavours. Be sure to buy the natural one.

Mustard Cabbage (Gaai Choy)

This is another of the innumerable Chinese greens. This cabbage has a spicy mustard flavour and transluscent green stems when cooked. Recognise it by thick green stems and more luminous green colour. These are basically three types:

- The long, thin one, with 5 or 6 per bunch – *bamboo mustard cabbage (chuk gaai choy)*.
- *Swatow mustard cabbage (daai gaai choy)* with thick, curved stems.
- *'Sow' cabbage* (jui la choi), which is the biggest. It has smooth-edged leaves, whereas the others have serrated-edged leaves. This one is the best to buy for cooking as a vegetable in its own right.

Nam Pla

A fish sauce strained from fermented fish. Too strong for some, but a necessary ingredient in much South-East Asian cooking. The ancient Romans used a similar sauce called liquamen.

Nori

Called *laver* in England and Ireland. A traditional food of many coastal peoples throughout the world. The Japanese have developed it as an item in cooking. You may know it as the wrapper in *nori sushi*. It has a delicious maritime flavour and is high in protein and minerals. It is available in sheets from Japanese, 'health' and natural food stores, toasted or untoasted. The toasted variety, *yaki nori*, can be used straight from the pack, whereas regular nori needs to be waved over a flame before use.

Nori Tsukedani

A delicious maritime-flavoured condiment made from nori seaweed, mirin, shoyu and rice malt (or black sugar). Used as a bread spread or in sauté.

Palm Sugar

Called *gula malacca, gula jawa* or *djaggery*, this sugar is simply the natural crystallised sugar of boiled down palm flower sap. It has a sweet smokey flavour and an unobtrusive sweetness. Used widely throughout Asia and India, it is very cooling and used to reduce blood pressure in some regional Asian medicine.

Polenta

Yellow, fine-grained cornmeal.

Rice Malt

A sweet syrup made by culturing rice and removing water by boiling or vacuum extraction. This is an authentic malt with a complex sweet flavour. It is not chemically manufactured as some malt extracts are.

Rice Vinegar

A traditional and authentic vinegar from naturally fermented rice. Once upon a time the European equivalent was malt vinegar — now just a memory. Rice vinegar has a very special flavour. It is without par as a seafood dip with shoyu and makes sushi like you'll never taste in a restaurant. Be sure to buy the natural one.

Sake

Alcoholic rice wine or spirit from Japan.

Sambal

A chilli paste or sauce. *Taucho sambal* is made from chillies and Indonesian 'miso' *(taocho); blachan sambal* is chillies and shrimp paste (blachan).

Sesame Purée

This is not *tahina*. Sesame purée is made from lightly roasted, unhulled sesame seeds and is available in Chinese stores, natural and 'health' food shops. The best one comes from Japan.

Shiitake Mushrooms

Often called *Chinese dried mushrooms*. Another characteristic Japanese flavour because, unlike in Chinese cooking, they are often used to make dashi or stock whose flavour forms a specific backdrop to many Japanese meals.

Star Anise

A star-shaped seed pod which has the true liquorice flavour, which may surprise you to know doesn't come from liquorice at all, but from anise, especially star anise. Available in Asian stores and some natural food stores.

Szechwan Red Pepper

A spicy, aromatic red peppercorn, available in Chinese stores. Characteristic in the cooking of Szechwan province (China), but also an exotic and well-liked flavour to experiment with. Is particularly good with miso.

Takuan
Daikon radish which is naturally pickled with rice bran and sea salt. A beautiful flavour and strong digestive. Usually eaten after meals in Japan. There are many Chinese equivalents, but takuan is a Japanese style.

Tamarind
A sour fruit from South-East Asia and North Australia. Used extensively in cooking. Comes as a block of compressed black fruit. May be called *imli* or *asam jawa*.

Tumeric
The same family as Laos and ginger, but bright yellow. Try to buy small dried roots and grate them on the nutmeg grater for use. This tastes much better than the packaged powder.

Umeboshi
Extraordinary pickled 'plums' from Japan. The ultimate for most stomach ills and a necessary fellow traveller in those countries where you may contract digestive-tract invaders. They lend a cleansing sharpness to dips and dressings and *have* to be tried on fresh boiled sweet corn.

Unbleached White Flour
Usually, white flour is automatically sterilised and bleached as it comes from the roller mill. If this process is not carried out, we have a creamy white flour which, depending on the sifting, contains a proportion of germ and a little bran. It is white flour with flavour.

80 Per Cent Flour
Wholemeal is 100 per cent of the wheat. If we sift this to remove the coarse bran, the flour is approximately 80 per cent 'wholemeal', with 20 per cent removed as bran. A good cake flour.

Wakame
A green sea vegetable from Japan. Used in salads and miso soup.

NOTE TO READERS

In the recipes included in this book, the following annotation has been used:

t = teaspoon
T = tablespoon

APPETIZERS & DIPS

STUFFED AGÉ (SERVES 6-8)

FRIED BEAN CURD CUT IN CUBES OR TRIANGLES, CHINESE
   STYLE (SEE INGREDIENTS & TECHNIQUES)

BOILING WATER

250 g FISH MINCED

1 CUP FRESH BASIL LEAVES

1 t RICE VINEGAR OR 1 T LIME JUICE

1 T NAM PLA (FISH SAUCE)

1 CHILLI SEEDED

1 CLOVE GARLIC

Begin by toasting the garlic and chilli. Do not peel. Insert a skewer (satay stick) and grill them over a flame (or element). The skin should blacken and you may think you've gone too far. Don't panic. The chilli and garlic and basil need to be pounded together with a mortar and pestle. Alternatively, chop as finely as possible and mix all ingredients, or use a food processor for the whole thing. Make a slit along one face of the bean curd – the diagonal if using the triangle shape – and place 1 T or so of the filling in each. Arrange in a steamer, preferably a Chinese bamboo one, and steam for 30 minutes. This filling can be simplified by using only minced fish filling if desired, or by including only the basil and vinegar flavour.

STUFFED AGÉ SALAD

10-12 CUBES BEAN CURD FRIED

2 BUNCHES (ABOUT 20) RADISHES WASHED

1 T DARK SESAME OIL

1 t OF EITHER HONEY OR BLACK SUGAR, OR
   1 T MALTOSE OR RICE MALT

1 T RICE VINEGAR

2 T TAMARI

Smash the radishes with the flat of a cleaver, or in a mortar and pestle or with a rolling pin, or briefly in a food processor. They should be smashed, not annihilated. Mix the condiments together and thoroughly combine with radishes. Slit the top of each bean curd cube, clear the pouch with a chopstick and place filling inside. Seal the top with a toothpick.

STUFFED TOFU

FRESH TOFU, DRAINED AND PATTED DRY

500 g FRESH ABALONE TRIMMED*

250 g FISH FILLETS

1 SPRING ONION STALK

WHITE PEPPER (OPTIONAL)

Chop onion finely and mince, grind or process together with fish and abalone. Add a little white pepper if desired. Form mixture into 12 balls to fill 12 triangles. Cut each cake about 20 cm thick and then diagonally in half into triangles. On the diagonal cut, make a slit in the face but don't cut through the triangle. Carefully insert the ball into each slit, flattening to fit. Heat oil. Deep fry until golden. Serve with chilli sauce and a vinegar-shoyu dip.

* The abalone should be trimmed of all black edges. It needs to be pounded or diced very finely before mincing or processing. You can substitute fish or prawns for the abalone, or puréed chickpeas, chopped peanuts and natto miso.

TEMPEH PÂTÉ (1)

½ BLOCK TEMPEH

2 T KECHAP MANIS

1 T LAOS

1 CUP BOILED OR ROASTED PEANUTS OR MACADAMIA NUTS

2 T LIME OR LEMON JUICE

Dice tempeh finely. Deep fry briefly till golden. Drain on paper towel. Steam or parboil tempeh if you prefer. Place all ingredients together and pound, purée or process. Add 1-2 roasted chillies or to taste, and 2 cloves roasted garlic for a fiery version. Serve with slices or wedges of raw carrot for dipping.

TEMPEH PÂTÉ (2)

½ BLOCK TEMPEH

2 T PREPARED MUSTARD (HOT OR MILD)

½ CUP ALMONDS TOASTED

3 T WHITE MISO

2 T FRESH CORIANDER FINELY CHOPPED

Proceed as above.

CHICK PEA PÂTÉ

5 CUPS CHICK PEAS COOKED WITH
   15 cm KOMBU SEAWEED

1 CUP FLAT-LEAF (CONTINENTAL) PARSLEY CHOPPED

JUICE OF 1 LEMON

½ CUP WHITE MISO

FRESH BLACK PEPPER (LOTS)

2 T SESAME PURÉE OR 3 T TAHINA OR ALMOND PASTE

1 T VIRGIN OLIVE OIL

2 T CORIANDER SEEDS FRESHLY GROUND

Pureé, pound or process all ingredients. Use some chick pea cooking liquid if thinning out is necessary, but keep mixture as thick as possible. Serve with thin crisps of deep-fried tempeh or toasted sourdough bread fingers.

BUDDHA 'CHICKEN'

227 g (8 oz) PACKET BEAN CURD STICKS (YUBA STICKS)

4 T LIGHT SESAME OIL

3 T SHOYU

2 T NATTO MISO

3 t PALM SUGAR OR BLACK SUGAR OR HONEY OR MALTOSE

1 T DARK SESAME OIL

2 STAR ANISEED

1 T MIRIN

Break the bean curd sticks in half and soak for 4 hours in water with a pinch of bicarb of soda. Place a plate on top to prevent floating. Rinse, soak in plain water for ½ hour or until being used. Pat dry on a towel before frying. Smash the aniseed and mix with shoyu. Soak for ½ hour. Heat light sesame oil in wok. When hot, add bean curd sticks and stir fry rapidly. Drain shoyu, discard aniseed. Add with remaining ingredients, except dark sesame oil, mix and fry over low heat for 10 minutes with lid on. Add dark sesame oil. Remove to a bowl and allow to cool. Now, to form a sausage shape, use a tea towel, squeeze the bean sticks and line them up a few inches from the edge of the cloth. Roll cloth over the sticks and then roll it up. Tie tightly with string all the way along, criss-crossing. Steam for 40 minutes. Refrigerate 2-4 hours. Unwrap, slice and serve as an appetizer or part of a meal.

TEMPEH CURRY PUFFS

Pastry:

2 CUPS UNBLEACHED WHITE OR 80% FLOUR

1/2 CUP CHICK PEA FLOUR (BESAN)

1/2-3/4 CUP SOY OR COCONUT MILK

SALT

Tempeh filling:

3 T PEANUT OIL

1 BLOCK TEMPEH FINELY DICED

1 ONION FINELY CHOPPED

1 CUP POTATO OR PUMPKIN PRE-BOILED AND DICED

2 t LAOS POWDER

3 t ROASTED OR DARK CURRY POWDER

1-3 FRESH CHILLIES SEEDED AND CHOPPED FINE (OPTIONAL)

2 T SHOYU OR KECHAP MANIS

Pastry: *Add salt to flours and mix well. Add liquid, starting at 1/2 cup. Coconut milk, may require 3/4 cup depending on how thick it is. Stir in with a wooden spoon, then knead with your hands till smooth. Roll out into circles 70 cm across. These should be thin. Dust each with a little flour, stack and cover with a cloth.*

Filling: *Heat oil and add onions. Fry over high heat briefly. Add tempeh and remaining ingredients except potatoes (or pumpkin). Fry hard for 3 minutes. Add potatoes or pumpkin and mix well. Place a spoonful of the filling in the centre of each round. Fold into a semicircle. Press down edges. Deep fry until golden, or bake.*

STUFFED MUSHROOMS

Choose even sized mushrooms. Remove stalk completely. Briefly rinse and pat dry. Steam for 5 minutes. Spoon Satay Peanut Sauce or Tempeh Pâté into cavities. Grill until toasted. You won't be able to make enough of these.

GUYANA TEMPEH DIP

3 LARGE TOMATOES PEELED AND SEEDED

3 FRESH RED CHILLIES OR 2-3 T PREPARED CHILLI
   (SEE GLOSSARY)

2 t MUSTARD POWDER

2 T DARK MISO

FRESH BLACK PEPPER (LOTS)

1 ONION FINELY DICED OR MINCED

2 CLOVES GARLIC CRUSHED

4 T CIDER OR RICE VINEGAR

1 T FRESH BASIL CHOPPED

1 T FRESH CORIANDER CHOPPED

1 t FRESH THYME CHOPPED

1 BLOCK TEMPEH

Put all the ingredients except tempeh and herbs in a saucepan. Simmer for 1 hour. Add basil, coriander and thyme. Slice straight through the tempeh so as to have long ovals. Deep fry until crisp. Dip into sauce.

SUSHI WITH TEMPEH

Rice must be cooked so it isn't gluggy. I use short-grain brown rice for this, but Thai jasmine rice is also excellent (refer Ingredients & Techniques)

3 CUPS HOT RICE FRESHLY COOKED

1 T RICE VINEGAR

1 t HONEY

½ BLOCK TEMPEH

Place rice in a flat bowl or large plate. Mix honey with vinegar and add to rice. Stir rapidly with a wooden paddle and lift the rice so as not to form clumps. Cover with a cloth. Cut tempeh into approximately 60 mm thin strips. Drop into boiling water to blanch for 30 seconds. Cool. Dip in shoyu.

To assemble: *With moist hands form rice into regular fingers. Place tempeh atop. Garnish with toasted sesame seeds or namprik.*

SUSHI WITH TOFU

Rice as for previous recipe.

Tofu:

Cut into 60 mm slices to go on your finger of sushi rice. Drop into boiling water for 30 seconds. Rinse, drain, cool and pat dry. Alternatively, if tofu is very fresh, simply rinse and pat dry. Cut thin (10 cm) strips of toasted (yaki) nori. Place tofu on top of rice finger. Carefully wrap two strips of nori onto each finger. Moisten if necessary. Place a piece of red pickled ginger on top.

Another way: Using Chinese-style fried tofu cubes, slit the top ready for stuffing and poke a hole with a chopstick.

Marinade:

1 CUP STOCK (SEE INGREDIENTS & TECHNIQUES)

3 T SHOYU

1 T MIRIN

Bring some water to the boil and drop bean curd in to remove excess oil. Then place in cold water and slowly reboil. Drain. Heat marinade. Drop bean curd in and leave overnight. Squeeze gently and fill each with sushi rice. Garnish with red pickled ginger.

POPIA (MAKES 6-8 ROLLS)

These are a Singaporean specialty – steamed rolls. Special paper-thin skins are made for popia from a sticky wheat dough slapped on a hot iron griddle. Regular spring roll skins are satisfactory. If you are adept at making thin crêpes, use them instead of the spring roll skins, or, if you have time, make the skins as directed below.

1 PKT MEDIUM-SIZE SPRING ROLL SKINS

OR

To make skins:

3 CUPS UNBLEACHED WHITE OR 80% WHOLEMEAL FLOUR

2-2¼ CUPS BOILING WATER

Mix flour with boiling water using a wooden spoon until handleable. Knead till smooth, adding more flour if necessary. Rest for 30 minutes while you make the filling. Roll up regular sized balls of the pastry. These must be smooth. When this is completed, roll each out into a very thin round. Use as directed for popia. May also be deep fried.

Filling

LETTUCE

2 CUPS DAIKON RADISH COARSELY GRATED (USE CARROT OR
TURNIP IF DAIKON ISN'T AVAILABLE)

6 AGÉ TOFU CUBES (CHINESE ONES) THINLY SLICED

1 CUP BEAN SPROUTS

1 t BLACHAN (PRAWN PASTE) OR USE 1 T DARK OR NATTO MISO

2 CLOVES GARLIC FINELY CHOPPED

3 T RICE MALT OR MALTOSE

CHILLI PASTE OR 1-3 FRESH CHILLIES

2 T TAMARI

3 T PEANUT OR LIGHT SESAME OIL

SQUEEZE LEMON OR LIME JUICE

Heat oil and add garlic and blachan. Stir and mix well. Add chillies (diced if using fresh) and maltose. Mix well. Add daikon and tamari, turn down heat and simmer for 5 minutes. Drain well. Cool. Spread out skin. Spread lettuce and place a line of daikon mixture on it. Cover with tofu slices, bean sprouts, lemon juice and chilli sauce if desired. Roll up, folding ends in, and steam for 15 minutes. In case the rolls stick to your steamer, line it with lettuce leaves and place rolls on top.

POTTED PEAS

500 g FRESH GREEN PEAS

2 T MINT FINELY CHOPPED

3 T RICE OR CIDER VINEGAR

3 T MIRIN

1 T WHITE MISO

2 T AGAR FLAKES

1 T LEMON OR LIME JUICE

1 T RICE MALT OR MALTOSE OR 1 t HONEY

Place peas in 3½ cups cold water to cover and bring to the boil. Simmer for 5 minutes. Strain, put peas in cold water and reserve stock. Add agar to stock and place on medium heat. Stir until agar is completely dissolved. Add remaining ingredients. Cook for 5 minutes. Add peas. Pour into one large or several small pots or decorative moulds. Chill. Another method is to purée the peas before they are added at the last stage. Slice into wedges, serve with dry discuits, deep-fried tempeh sticks or as a side dish, especially on Christmas Day.

SAMOSA

I can't resist them and nobody else seems to be able to either!

Skins:

2 CUPS UNBLEACHED OR 80% WHOLEMEAL FLOUR

SALT TO TASTE

2 T GHEE OR PEANUT OIL

4 T SOYOGHURT OR YOGHURT

Add salt to flour. Blend in ghee or oil with a wooden spoon until well distributed. Add yoghurt and knead until a smooth dough is formed. Use more yoghurt if necessary. Roll into smooth balls 40 cm in diameter. Cover.

Filling:

1 BLOCK FIRM TOFU (450-500 g) in small dices

1 T PEANUT OIL OR GHEE

2 T GINGER FINELY DICED

1 CUP FRESH GREEN PEAS

1 t CORIANDER POWDER FRESHLY GROUND

1 CHILLI DICED OR ½ t CHILLI POWDER

¼ t WHOLE BLACK MUSTARD SEEDS

¼ t GARAM MASALA

¼ t TUMERIC POWDER

SALT TO TASTE

Heat oil and sauté ginger, mustard seeds, chilli, masala, tumeric and coriander for 2-3 minutes over medium heat. Add peas. Cover and cook slowly for 10 minutes. Add tofu and salt. Mix and cook 2 minutes more strain the mixture. Roll out pastry into thin rounds. Cut in half. Place some filling on each half and fold over to form a cone or triangle. Seal edges. Deep fry and serve with Miso Chutney. Alternatively, steam them for 30 minutes or bake for 20 minutes at 200°C (400°F).

YUBA-TEMPEH (SERVES 6)

East meets East.
You will need dried bean curd sheets (yuba), not sticks, for these. The more pliable (i.e. less brittle) the better. Soak yuba in warm water for 1-2 minutes. Drain and place on a towel on the chopping board. Spread it out and cut into 10 cm squares. Cover with a moist towel.

Tempeh

½-¾ BLOCK FRESH TEMPEH CUT INTO STICKS
   (50 x 10 x 10 mm)

Sauce:

5 T TAMARI

3 T RICE VINEGAR

4 T MALTOSE, RICE MALT OR 2 T HONEY

1 t RED SZECHWAN PEPPER CRUSHED

1 CLOVE GARLIC MINCED OR 1 t GINGER FINELY GRATED

Mix all sauce ingredients in a saucepan and simmer for 15 minutes. Place tempeh inside yuba square. Roll up, folding in ends. To seal, slip a slice of hollowed-out cucumber or carrot over each end to seal or use flour paste or a toothpick. Deep fry until golden (careful, they splatter!) or bake at 180°C (350°F) for 20 minutes. Arrange on plate, pour over sauce and serve.

TEMPEH TEMPLE BALLS

These are cubes of steamed, baked or deep-fried tempeh, marinated in chutney, encased in glutinous rice, rolled in sesame seeds and deep fried, steamed or baked – deep frying works best. Brown rice is suitable if no glutinous is available.

Glutinous rice: *To cook rice, refer to Ingredients & Techniques. Allow rice to cool. Turn rice into a bowl and knead with wet hands until it's sticky and well broken up.*

Chutney:

1 CUP DATES

1 T GINGER FINELY GRATED

1 t MANDARIN OR ORANGE RIND FINELY CHOPPED

1 t CORIANDER SEED FRESHLY GROUND

2 T DARK MISO

1 CUP WATER

PINCH CAYENNE (OPTIONAL)

Chop dates finely. Add to water with other ingredients, being sure to dissolve miso thoroughly. Slowly bring to the boil then simmer until dates are very soft.

Tempeh: *Deep fry in cubes roughly 20 mm square. Alternatively, steam or bake it. Then marinate in chutney at least 1 hour.*

To assemble: *With wet hands form glutinous rice into balls. Make a hollow in the ball. Insert the tempeh cube, making sure to get plenty of chutney inside. Seal and smooth with wet hands. Roll in sesame seeds and deep fry. Alternatively, steam for 30 minutes or bake for 30 minutes at 200°C (400°F). Serve with vinegared cucumber.*

SESAME TOFU

3 CUPS COLD WATER OR KOMBU STOCK

SALT TO TASTE

1½ t MIRIN

4 T (HEAPED) KUZU OR 5 T ARROWROOT (KUZU GIVES THE
  BEST TEXTURE)

4 T TAHINA OR 3 T SESAME PURÉE

3 T SOYMILK

Dissolve kuzu thoroughly in water. Add salt and mirin. Slowly bring to the boil, then simmer, stirring constantly until thick and clear. Remove 3 T and dissolve thoroughly with soymilk, tahina or sesame purée. A whisk is useful. Mix back in and stir thoroughly until creamy. Simmer over very low heat for 5 minutes. Pour into a rinsed glass dish. The 'tofu' should be about 10 cm thick. Cool. Refrigerate. When firm cut into bite-size squares.

Dipping sauce:

3 T SHOYU

1 T LEMON JUICE OR 2 T LIME JUICE

This is a mouth-melting entreé or appetizer, best eaten with chopsticks. It is also delicious using almond butter instead of tahina, in which case, change the dip to shoyu and mustard. Freshly toasted ground sesame seeds make this dish superlative. Use 6 T ground seeds and cream with a little water or soymilk before adding in place of tahina.

TOFU DIPS – These also double as salad dressings

TOFU-MAYO:

250 g TOFU

2 T PREPARED MUSTARD

1 T ALMOND BUTTER

1 T LEMON JUICE

1 T VIRGIN OLIVE OIL

¼ CUP WATER OR STOCK

Pound, purée or process all ingredients. Chill. Serve with black bread and slivers of toasted nori.

DILL TOFU:

250 g TOFU MASHED

3-4 GHERKINS PICKLED IN BRINE AND CHOPPED

1 T CHOPPED FRESH DILL

2 t HONEY

1 T LEMON JUICE

Pound, purée or process all ingredients till smooth. Chill. Serve with corn chips, rice crackers or black bread fingers.

PINKO TOFU:

250 g TOFU MASHED

1 MEDIUM BEETROOT COOKED, SLICED AND SOUSED WITH
   CIDER VINEGAR

½ t CARAWAY SEEDS CRUSHED

1 SPRING ONION STALK CHOPPED

Pound, purée or process all ingredients until smooth. Use apple juice to thin if necessary. Serve with sashimi or sushi or mix with blanched, chilled vegetables as a salad.

BURMESE-STYLE TEMPEH DIP

The Burmese use a different soya bean cake but tempeh works well.

½ BLOCK OF TEMPEH IN THIN STRIPS

4 MEDIUM TOMATOES WELL RIPENED

2 CLOVES GARLIC UNPEELED

2 SMALL BROWN ONIONS

4 GREEN CHILLIES SEEDED

SALT TO TASTE (1 t)

The tempeh is grilled or baked until golden brown and crisp. The garlic, onions, chillies are baked or grilled until soft. Pierce with a pin to prevent explosions. The tomato can also be baked, grilled or seared. When soft and almost melting, the skin can be removed easily and discarded. Pound, purée or process all ingredients to a paste. Serve with vegetable sticks or crisped fingers of sourdough bread, or eat as a side dish with rice or noodles and vegetables.

This sauce or dip would be called Nam Prik in Thailand, meaning a seared-pounded chilli sauce. Each region has its own specialty, and often every cook does as well. The most interesting one uses dried grasshoppers called Maeng-Da instead of tempeh or shrimp paste. It is very good.

TOFU DIPS – These also double as salad dressings

TOFU-MAYO:

250 g TOFU

2 T PREPARED MUSTARD

1 T ALMOND BUTTER

1 T LEMON JUICE

1 T VIRGIN OLIVE OIL

¼ CUP WATER OR STOCK

Pound, purée or process all ingredients. Chill. Serve with black bread and slivers of toasted nori.

DILL TOFU:

250 g TOFU MASHED

3-4 GHERKINS PICKLED IN BRINE AND CHOPPED

1 T CHOPPED FRESH DILL

2 t HONEY OR APPLE JUICE CONCENTRATE

1 T LEMON JUICE

Pound, purée or process all ingredients till smooth. Chill. Serve with corn chips, rice crackers or black bread fingers.

PINKO TOFU:

250 g TOFU MASHED

1 MEDIUM BEETROOT COOKED, SLICED AND SOUSED WITH
   CIDER VINEGAR

½ t CARAWAY SEEDS CRUSHED

1 SPRING ONION STALK CHOPPED

Pound, purée or process all ingredients until smooth. Use apple juice to thin if necessary. Serve with sashimi or sushi.

'SAFFRON' TOFU

250 g TOFU MASHED

2 t TUMERIC POWDER

2 T LEMON OR LIME JUICE

1 SMALL CUCUMBER PEELED AND DICED

Pound, purée or process all ingredients until smooth. Thin with soymilk if necessary. Serve with pappadams.

LIME TOFU:

250 g TOFU MASHED

1/2 BUNCH PARSLEY

4 T LIME JUICE

2 T TAHINA OR 1 T SESAME PURÉE

1 T WHITE MISO

Chop the parsley very finely. Place in a cheesecloth and squeeze juice into tofu. Dip in lime juice, wring and squeeze out more until tofu is pale green. Discard parsley. Purée remaining ingredients and serve with seafood.

TOMATOFU DIP:

250 g TOFU MASHED

2 MEDIUM RIPE TOMATOES SKINNED AND SEEDED
   OR 2 T TOMATO PASTE

1 t GINGER FINELY GRATED

SALT TO TASTE

FRESHLY GROUND BLACK PEPPER (LOTS)

1 T CIDER VINEGAR (PICKLE VINEGAR FROM ONIONS
   WORKS WELL HERE)

Mix cider vinegar and ginger and leave for 5 minutes. Pound, purée or process all ingredients until smooth. Serve with deep-fried rice balls.

MARINADES & ACCOMPANIMENTS

TERIYAKI

6 T DARK MISO

3 T MIRIN

1 T GINGER FINELY GRATED

1 T GARLIC CRUSHED

1 T HONEY OR 2 T RICE MALT OR MALTOSE

1 CUP WATER

Mix all ingredients thoroughly. Slowly heat until almost boiling. Cool. Preferably prepare 1 week before use. Use on fresh, grilled or barbecued tofu, fish or tempeh.

HOT 'TERIYAKI'

Add from 1 t – 1 T finely minced chilli to above sauce.

DRUNKEN TERIYAKI

Add 3 T saké instead of mirin to original recipe.

SZECHWAN

Add 1 t ground red Szechwan pepper to original recipe with 1 t chopped fresh chilli.

HOT & SOUR

Add 2 T rice vinegar and 1 t fresh chilli to original recipe.

JAVA

Omit mirin. Substitute 1 T Laos powder for ginger, use 3 T palm sugar (gula malacca or gula jawa) and add 1 t finely chopped chilli.

NAM PRIK

Roughly translated from Thai as 'chilli water' or 'chilli sauce'. There are innumerable variations on this theme.

1 T DRIED SHRIMP

4 CLOVES GARLIC

3 t BLACHAN

3-7 RED CHILLIES

JUICE OF 2 LIMES

2 T PALM SUGAR

2 T NAM PLA

1-3 CUPS FRESH BASIL LEAVES

Pound, purée or process dried shrimp until a powder. Wrap blachan in foil, prick the chillies, garlic and onions and roast all in the oven or over a flame. The garlic should be slightly charred and oily. Blachan needs about 3 minutes over a flame or 10 minutes in a 200°C (400°F) oven. Purée all ingredients. Serve as an accompaniment to deep-fried tempeh or tofu, or with kebabs or seafood.

- Substitute ½ cup mint or fresh coriander for basil.
- Omit basil leaves in original recipe.
- Substitute 1 T tamarind paste for lime juice.
- Vegetarian: Substitute 3 T dark or light miso for blachan. Omit shrimps and nam pla. Add ½ cup toasted crumbed nori and 1 cup soaked wakame seaweed and use 2 T lime or lemon juice, or tamarind paste.

MARINADES
FOR GRILLING:

1. ½ CUP LOW SALT SHOYU
   ¼ CUP WATER
   4 T MIRIN

2. ADD 2 T RICE VINEGAR OR LEMON JUICE OR

3. ADD 1 T FINELY GRATED GINGER OR

4. ADD 1 t FINELY MINCED GARLIC OR

5. ADD 1-2 T AUTHENTIC CURRY POWDER (CHANNA MASALA IS GOOD) OR

6. ADD 1 t GROUND SZECHWAN PEPPER OR

7. 3 T PREPARED MUSTARD OR

8. 1 t DARK SESAME OIL OR

9. ORANGE ZEST OR

10. 1 t CUMMIN POWDER
    1 t CORIANDER POWDER
    1 t CRUSHED GARLIC

SPECIAL TEMPEH MARINADE

¼ CUP LOW SALT SHOYU

¼ CUP KECHAP MANIS

¼ CUP WATER

1 t LAOS POWDER

1 t CRUSHED GARLIC

'DEVILLED' MARINADE

4 T TAMARI

1 T CIDER VINEGAR

1 t GINGER FINELY GRATED

1 t GARLIC FINELY CHOPPED

2 T RICE MALT OR 1 T HONEY

1 T TOMATO PASTE

1 T DARK SESAME OIL

2 T MIRIN OR SAKÉ

1 t CUMMIN POWDER

ANISE MARINADE

1/2 CUP TAMARI

ZEST FROM ONE ORANGE

4 WHOLE STAR ANISE

1 T MIRIN

1/4 CUP WATER

Slowly heat, but don't boil, for 2-3 minutes. Cool. This is best made 1 week in advance

SWEET & SOUR

1/4 CUP SHOYU

1/4 CUP RICE VINEGAR

1/2 CUP RICE MALT OR MALTOSE

OR OMIT THE SHOYU

OR ADD CHILLI

Mix well.

SOUPS

MISO SOUPS

WITH TOFU:

1 BAY LEAF

2 LITRES WATER

20 cm WAKAME SEAWEED

250 g TOFU DICED

1 MEDIUM ONION THINLY SLICED

1 t GINGER FINELY GRATED

1 MEDIUM CARROT DICED

½ MEDIUM MUSTARD CABBAGE OR CHINESE CABBAGE CUT
   ACROSS THE STEM INTO STRIPS

DARK RICE (GENMAI) OR BARLEY (MUGI) MISO

Add bay leaf and carrot to cold water. Bring to boil. Add onion and wakame. Simmer 10 minutes covered. Bring to gentle boil again and add tofu and greens. Allow to reboil, simmer covered for 5 minutes. Add ginger. Take a cup of broth from the saucepan, dissolve miso in it with a wooden spoon. Return to soup, let stand or simmer very slowly for 5 minutes. The amount of miso used depends on your taste (1 t per person is average). This soup serves 7-8, but is worth making in this quantity even for a single because it can be refrigerated and a small portion heated each morning for a hearty start to the day.

WITH CORN:

CORN COBS (1 PER PERSON)

1 STICK KOMBU

1 OR 2 CARROTS SLICED

4 MEDIUM TOMATOES PEELED AND FINELY CHOPPED

2 ONIONS DICED

⅓ BUNCH BOK CHOY OR OTHER CHINESE CABBAGE SLICED

1 t GINGER FINELY GRATED

4 T MISO (GENMAI OR MUGI ARE BEST)

Boil the cobs in water with one stick of kombu seaweed. You will need about 1¾ litres corn stock. Remove corn from stock. Bring to a boil, add carrots, tomatoes and onions. Simmer rapidly for 10 minutes Add bok choy and ginger. Simmer rapidly for 2-3 minutes. Remove 1 cup of stock. Dissolve 4 T miso in it and return to soup. Simmer very slowly for 5 minutes.

Serve with corn, salad and sourdough bread for a perfect summer meal.

WHITE (SHIRO) MISO:

2 STICKS KOMBU SEAWEED

3 SHIITAKE MUSHROOMS

1 LITRE WATER

½ CUP WHITE (SHIRO) MISO

1 DAIKON

½ CUP LEEK FINELY SLICED IN ROUNDS

6 CUBES AGÉ TOFU SLICED AND WASHED IN HOT WATER

CHINESE CABBAGE (GREEN PART) THINLY SLICED,
   OR SILVER BEET 2-3 HANDFULLS

1 SPRING ONION

Rinse shiitake. Soak the kombu and shiitake in cold water for 10 minutes. Slowly bring to the boil, but don't boil. Simmer over low heat for 20 minutes. Strain the stock. Use kombu elsewhere. Reserve shiitake. Remove stalk and slice mushroom thinly, then add to stock with daikon. This should be sliced in half and then each half sliced to make semicircles. Bring stock to the boil, add leeks and tofu. Simmer for 2-3 minutes. Add greens. Remove 1 cup stock. Dissolve white miso in it and return to stock. Simmer 2-3 minutes. Serve garnished with thin slices of spring onion.

BASIC MISO:

Make a simple stock with a seasonal standby and simply add miso. For example, bring 1¼ litres of water to the boil. Add 3 cups washed, chopped watercress. Allow to reboil, then simmer, covered, for 5 minutes. Remove 1 cup stock, dissolve 3-4 T of brown miso in it and return to soup. Simmer 2-3 minutes. Add a little grated ginger or sliced spring onions if desired. Use asparagus instead of watercress and substitute 6 T white (shiro) miso. Garnish with red chillies thinly sliced.

LENTIL MISO:

1 ½ CUPS BROWN LENTILS WASHED

2 BAY LEAVES

4 LARGE TOMATOES SKINNED AND SEEDED

1 STICK CELERY OR FENNEL SLICED

1 ONION FINELY DICED

3 T OLIVE OIL

1 T CORIANDER SEED GROUND

3-4 T BARLEY MISO (MUGI)

½ CUP PARSLEY FINELY CHOPPED

Cook lentils in 3 times their volume of water with bay leaf until soft. Add 1 ¼ litres water. Bring slowly to the boil. Add finely diced tomatoes and fennel. Sauté onion in olive oil with coriander and parsley. When onion becomes clear, add to the lentils and mix well. Take out 1 cup stock and dissolve miso in it. Return to soup and simmer for 5 minutes.

KIDNEY BEAN AND MISO:

2 CUPS PUMPKIN CUBED

1 CUP CARROT DICED

1 CUP CELERY SLICED

2 ONIONS FINELY CHOPPED

2 BAY LEAVES

2 STICKS KOMBU OR WAKAME (ABOUT 120 cm)

4 T MISO (RED OR BROWN)

1 ½ LITRES WATER

This soup needs 1 ½ litres of water, so if you have any cooking liquid left from the kidney beans, incorporate it in the water quantity.

To cook beans: *Soak 300 g overnight. Discard water and wash beans. Place beans in 1 ½ litres of water with kombu or wakame and bay leaf. Bring to the boil, lid off, and simmer rapidly for 10 minutes. Then simmer slowly, lid off, for 30 minutes. The vegetables are added at this stage and allowed to simmer, lid on, for 30-40 minutes with the beans. Remove 1 cup stock, dissolve miso in it and return to soup. Simmer for 5 minutes. You can substitute either chick peas, lima beans or mung beans for kidney beans, and add ginger if desired.*

MISO CONGEE

1 ¼ LITRES COLD WATER

1 STICK KOMBU SEAWEED

5 CUPS COOKED BROWN RICE (OR MILLET, BARLEY ETC)

4 OR 5 SHIITAKE MUSHROOMS

¼ CUP SPRING ONION CHOPPED

1 CUBE AGÉ TOFU PER BOWL

4 T WHITE (SHIRO) MISO *OR* 3 T NATTO MISO

TAKUAN RADISH PICKLE

Add kombu and rinsed shiitake to 1 litre of cold water. Bring to the boil over high heat. Remove from heat before it boils. Simmer. Purée rice briefly with ½ litre water — but don't completely purée it. Add to soup, bring to near boil and then simmer for 30 minutes stir often. Slice shiitake if desired and return to soup. Remove 1 cup stock and cream with miso. Return to soup and simmer for 5 minutes. Garnish with sliced agé, spring onion and slivers of pickled daikon (takuan). Use hatcho miso in winter.

TO WARM THE COCKLES OF YOUR HEART

2 LITRES COLD WATER

1 CUP AZUKI BEANS WASHED

1 STICK KOMBU SEAWEED

2 CARROTS DICED

1 TURNIP DICED

1 PARSNIP DICED

1 CUP PUMPKIN CHOPPED

2 ONIONS FINELY DICED

1 CUP CABBAGE SLICED

4 T HATCHO MISO

1 T GINGER FINELY GRATED

Place beans in cold water with kombu. Bring to a boil and simmer for 30 minutes uncovered. Add the remaining vegetables except the onion and bring to the boil. Add onion, mix well, add ginger, mix in and simmer for 20 minutes. Remove 1 cup stock and dissolve miso in it thoroughly. Return to soup and simmer 10 minutes longer. This hearty 'peasant' soup is ungarnished and best served with huge chunks of sourdough bread or brown rice and accompanied by pickled daikon radish (takuan). To complete the warm up, finish the meal with hot saké.

CREAM OF MUSHROOM

You need a large pot, but the mushrooms cook down. This is best with field mushrooms.

1 kg MUSHROOMS
5 CLOVES GARLIC COARSELY CHOPPED
2 BAY LEAVES
1 t SEA SALT
4 T VIRGIN OLIVE OIL
1 t BASIL DRIED
A LOT OF BLACK PEPPER FRESHLY GROUND
5 CUPS CHICKEN STOCK OR USE VEGETABLE-SEAWEED STOCK (SEE *INGREDIENTS AND TECHNIQUES*)
1 T EACH GENMAI, MUGI AND HATCHO MISO, OR 3 T OF ANY ONE OF THEM
200 g TOFU AND A SQUEEZE OF LEMON JUICE

Heat the oil in a large saucepan over high heat. When it is very hot throw in garlic, bay leaves. Stir briefly and add all of the mushrooms suitably chopped (i.e. all about the same size). Stir over high heat for 5 minutes until the mushrooms are 'sweating'. Add 1 teaspoon of sea salt, cover and simmer rapidly for 15 minutes, stirring occasionally. The mushrooms should lose much liquid. Add the basil and chicken stock and rapidly return to the boil. Simmer for 5 minutes. Strain the soup and pound, purée or process the solids with a little of the liquid and the miso. Mix into the soup and simmer for 5 minutes before serving. Add black pepper. Purée tofu with a little lemon juice and water, Swirl 1 T into each soup and serve.

CREAM OF CHICKPEA SOUP

1¼ LITRES WATER

3 CUPS CHICKPEAS SOAKED OVERNIGHT AND RINSED

1 STICK KOMBU SEAWEED

4 T WHITE MISO OR 3 T LIGHT BROWN MISO

½ BUNCH PARSLEY (FLAT-LEAF VARIETY) FINELY CHOPPED

HANDFUL OF FRESH CORIANDER, TORN RATHER THAN CHOPPED

2 T LEMON JUICE

3 T TAHINA OR 2 T SESAME PURÉE OR VIRGIN OLIVE OIL

SALT TO TASTE

PAPRIKA

Place chickpeas in saucepan with water and kombu. Bring to a boil, lid off. Simmer rapidly for 20 minutes, skimming occasionally. Continue to simmer slowly, lid on, until beans are soft. Strain and purée the beans with parsley, miso, lemon, tahina. Use reserved stock until the required thickness is obtained for the soup. Reheat. Stir in fresh coriander and salt. Serve sprinkled with paprika.

BARLEY-MISO

1 CUP WHITE BEANS SUCH AS HARICOT OR BLACK-EYE

   (WHICH DO NOT REQUIRE SOAKING OVERNIGHT)

1½ LITRES WATER

¾ CUP BARLEY GRAIN WASHED

WAKAME SEAWEED

1 CUP PUMPKIN OR CABBAGE DICED

1 CUP SLICED CELERY OR FENNEL

1 ONION FINELY DICED

3-4 T BARLEY (MUGI) MISO

Place beans in water and bring to a boil. Simmer rapidly uncovered for 10 minutes. Cover and simmer slowly for 45 minutes. Add barley, seaweed and remaining vegetables. Bring to a boil. Simmer covered for 30 minutes. Remove 1 cup of stock. Dissolve miso in it and return to soup. Mix well.

TOMATO AND MISO

Hopefully you have luscious home-grown tomatoes or at least organic ones. Tomatoes ain't what they used to be when we rated their quality (i.e. flavour, size, texture, juice) over their marketability – (uniform size, ripen when picked green, tough skins that travel well, etc.). Who are these mad people trying to put all manifestations into some sort of mechanical order? Subvert the dominant paradigm! The tomatoes need to be skinned, so immerse in boiling water briefly and then peel skin off. This soup is better without the seeds, so scoop them out and strain them to get all the juice out. Discard seeds.

8-10 MEDIUM RIPE TOMATOES, PEELED AND SEEDED
   AND FINELY CHOPPED

3 T VIRGIN OLIVE OIL

5 CLOVES GARLIC CRUSHED

SALT TO TASTE

2-3 CUPS WATER, DEPENDING ON JUICINESS OF TOMATOES

1 T DARK OR HATCHO MISO OR 2 T TAMARI

2 T FRESH BASIL CHOPPED

1 t GINGER FINELY GRATED

1 RED CHILLI SEEDED AND CHOPPED VERY FINE

Heat oil. Add garlic, sauté briefly, add tomatoes, chilli and salt. Sauté for 5 minutes until bubbling. Add water and bring to a boil. Add ginger. Remove ½ cup stock, dissolve miso in it and return to soup. Simmer for 5 minutes. Purée if you like. Serve with fried noodles.

GREENS AND MISO

1 CUP TOFU MASHED

1½ LITRES WATER

1 STICK KOMBU SEAWEED

1 LARGE OR 2 MEDIUM/SMALL BOK CHOY (CHINESE CABBAGE)
   WASHED WELL AND CHOPPED INTO SLICES

1 t GINGER FINELY GRATED

3-4 T BROWN RICE (GENMAI) OR BARLEY (MUGI) MISO

Bring water and kombu to a boil. Remove kombu and add greens. Over high heat, bring to a boil again. Simmer with lid on for 5 minutes. Take out 1 cup stock. Dissolve miso in it with ginger and return to soup. Add tofu, stir well. Simmer for 5 minutes. Use spinach if Chinese greens are not available and add 2 T lemon juice with tofu.

BEANCURD AND COCONUT

1 BLOCK TOFU DICED

MILK FROM ONE COCONUT OR ONE TIN MADE UP TO 1½ LITRES
    WITH WATER (SEE *INGREDIENTS AND TECHNIQUES*)

2 t LAOS POWDER

2 WHOLE STALKS OF LEMONGRASS OR 1 T POWDER

½ CUP WHITE MISO

1 CHILLI SEEDED AND SLICED IN THIN ROUNDS

JUICE OF 1 LIME

2 T FRESH BASIL OR MINT FINELY CHOPPED, OR 1 T OF EACH

1 T NAM PLA

FRESH CORIANDER

Chop the lemongrass and bruise it with the flat of a cleaver. Add Laos and lemongrass to coconut milk and slowly bring to the boil while stirring gently. Simmer for 10 minutes. Strain off lemongrass. Add tofu. Remove 2 cups of stock and dissolve miso, simmer for 5 minutes. Add remaining ingredients and serve garnished with coriander. This is delicious with a bowl of Thai jasmine rice.

SOUP NOODLES

These are easy, one-pot meals. The separate ingredients, stock and noodles can be prepared beforehand, refrigerated and assembled easily. There are innumerable variations on the themes presented here.

TEMPURA SOBA (DEEP-FRIED VEGETABLES AND BUCKWHEAT
    NOODLES (SERVES 6)

PRIMARY STOCK (DASHI):

1 LITRE COLD WATER

2 STICKS KOMBU SEAWEED

3 T (HEAPED) DRIED BONITO FLAKES OR USE INSTANT DASHI
    (SEE *INGREDIENTS & TECHNIQUES*)

Add kombu to water. Slowly bring to the boil — it should take 10 minutes. When nearly boiling, add ½ cup cold water and bonito flakes. As soon as the water re-boils, turn down to simmer for 1 minute. Strain the stock.

To cook soba

Bring 1 ½ litres water to a rolling boil. Add 1 pkt (675 g) soba noodles. Stir once or twice with a chopstick. When a head of foam rises suddenly in the pot and threatens to envelop the stove, add 1 cup cold water. Do this 3 times. When the noodles finally re-boil, strain and run under cold water until cool. Reserve them in a strainer.

TO MAKE THE TEMPURA BATTER:

2 CUPS 80% WHOLEMEAL FLOUR

3 CUPS ICE WATER

OR

2 CUPS UNBLEACHED FLOUR

2 CUPS ICE WATER

2 EGG YOLKS

OIL FOR DEEP FRYING

Either way, don't mix the batter in advance. Have all of the ingredients and utensils ready. You need chopsticks or a fork to mix the batter.

VEGETABLES:

1 SPRING ONION STALK THINLY SLICED ON THE DIAGONAL

6 RECTANGULAR PIECES OF TOFU 10 mm THICK

6 ONION SLICES HELD TOGETHER BY A TOOTHPICK THROUGH
    EACH SLICE

6 SECTIONS OF MUSTARD CABBAGE STALK 60 cm LONG

Wash the greens and drain. Pat dry the tofu. Dust all the vegetables with 80% wholemeal flour. Heat the oil and add a little batter. If it drops to the bottom of the oil and immediately rises to the top sizzling, the oil is ready.

Make the batter by placing the egg yolk in a small bowl, mixing briefly with chopsticks or fork. Add the water, mixing a little, and then the flour. Mix very briefly, so you think it's not mixed properly. That's it. Dip vegetables in batter. Immediately start your deep frying, 6 pieces at a time, until just golden brown. Drain on paper.

Assemble 6 bowls. Place the noodles under hot or boiling water to warm. Heat the stock, flavour with shoyu (2-3 T). Arrange noodles in bowl. Place tempura on noodles. Ladle in stock. Garnish with spring onions. Mix 2 T shoyu, ½ T ginger juice (squeezed from finely grated ginger) and 1 T hot tea as a dip for the deep-fried items.

Substitute udon for soba and try prawns, fish or other vegetables deep fried.

LAKSA (SERVES 6)

This is Penang-style noodles, a rich coconut broth with vegetables and/or seafood — and it's hot!

1 LITRE CHICKEN, FISH OR VEGETABLE STOCK

1 T (HEAPED) LAOS POWDER

3 GARLIC CLOVES CHOPPED

1 ONION DICED

1-5 CHILLIES FINELY CHOPPED

2 T BLACHAN

4 T PEANUT OIL

½ LITRE COCONUT MILK

3 T TAMARI

1 T GULA MALACCA (PALM SUGAR)

1 T PENANG-STYLE CURRY POWDER (OR MALAYSIAN)

SALT TO TASTE

12 CUBES AGÉ TOFU

BEAN SPROUTS

BROWN RICE NOODLES (GENMAI UDON) OR 1 PKT CHINESE
   NOODLES

Chinese noodles need only to be plunged in boiling water and boiled once, then strained and washed quickly. Cook the udon according to previous recipe.

Add the Laos powder to the chicken or fish stock. Slowly bring to the boil. Pound, purée or process garlic, onion, chilli and blachan to a paste. Heat oil and fry this paste over medium heat for about 10 minutes. Add curry powder and stir in. Add this to stock with coconut milk, tamari and gula. Salt to taste. Bring to a boil slowly while stirring. Simmer for 10 minutes. Place noodles in a bowl. Put agé tofu and bean sprouts on top and ladle in the broth. If using seafood — prawns, fish pieces, squid, abalone, fish balls or sliced fish cake — add with the coconut milk.

TOFU-EGG NOODLES (SERVES 6)

Quick and nourishing.

1 ½ LITRES COLD WATER

2 STICKS KOMBU SEAWEED

5 SHIITAKE MUSHROOMS RINSED

1 T MIRIN

1 CARROT SLICED

2 CUPS CHINESE GREENS SLICED

1 EGG PER PERSON

SHOYU TO TASTE

SPRING ONION SLICED TO GARNISH

BLACK PEPPER

DARK SESAME OIL

1 PKT UDON OR CHINESE NOODLES

1 BLOCK FRESH TOFU SLICED OR DICED

BEAN SPROUTS

Cook noodles, rinse under cold water and drain. Add kombu to water with shiitake and carrot. Bring to a boil slowly. Simmer for 15 minutes till shiitake mushrooms are soft. Remove kombu. Add Chinese greens and mirin. Place some noodles in each bowl. Add ¼ teaspoon dark sesame oil to each, then add tofu, bean sprouts, spring onions and 1 T shoyu or to taste. Slice the shiitake and share them between the bowls. Add hot broth. If you don't mind eggs only slightly done, break into the bowl just after adding stock. Alternatively, separately poach the eggs in shallow water and put on top of noodles. Grind lots of black pepper on top.

A shortcut is to use a pre-made dashi (stock). Make sure it is a natural one. For a vegetarian version, omit egg and add some agé tofu slices.

FISH-BALL NOODLES WITH TOFU (SERVES 6)

Fish balls are very easy to make if you can buy minced fish or make your own at home. Alternatively, scrape the flesh from a fish with a metal spoon. Chop it finely and pound. At this stage, chilli or spring onion can be pounded with the fish flesh for a different type of fish ball. Form the mince into small balls with moist hands. Bring water and 1 stick kombu seaweed to the boil. Drop the fish balls in. They will rise to the surface and are cooked 1 minute later. Drain the balls and reserve stock, keeping hot.

For this dish use 500 g fish flesh and mince or mix with . . .

1 SPRING ONION STALK

A LITTLE WHITE PEPPER

1½ LITRES FISHBALL STOCK

12 CUBES AGÉ TOFU

SPRING ONION AND CORIANDER TO GARNISH

1½ CUPS CHINESE CABBAGE (BOK CHOY) OR OTHER GREENS
   THINLY SLICED

SHOYU TO TASTE

CHILLI IF DESIRED

BEAN SPROUTS

Add greens to hot stock. Place some noodles in bowl with 3 fishballs, two sliced agé cubes, bean sprouts. Pour on stock and greens and sprinkle with spring onions and coriander, and shoyu (1 T per bowl usually). Fresh chilli can be sprinkled on or chilli condiment such as tao tjo sambal (taucho) mixed in by each person.

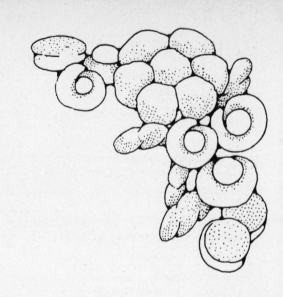

MAIN COURSES

GRILLED TOFU

Probably the easiest!

MARINADE:

½ CUP SHOYU (PREFERABLY LOW SALT)

4 T MIRIN

4 T WATER

ADD 4 T RICE VINEGAR FOR A DIFFERENT MARINADE (SEE ALSO
   MARINADES AND ACCOMPANIMENTS)

1 BLOCK TOFU (450-500 g)

*Slice tofu into rectangles 10mm thick. Pat dry and place in marinade at least ½
hour. Two hours is better. Brush dark sesame oil in a metal tray and arrange tofu
on it to suit your griller. Grill until spattered with brown high spots. Turn over
and repeat. Serve with stir-fried vegetables and brown rice, or slip between slices
of bread and salad.*

*The tofu can be grilled without using a marinade. Serve with tao-tjo (taucho)
sambal or a miso sauce such as:*

5 T DARK MISO (GENMAI, MUGI)

1 HEAPED t GARLIC FINELY CRUSHED

1 HEAPED t GINGER FINELY GRATED

2 T RICE MALT, MALTOSE OR 2 t HONEY

1 T RICE VINEGAR

3 T WATER

Mix well and place in a saucepan. Simmer over low heat for 10 minutes.

SAFFRON RICE WITH TEMPEH (SERVES 6-8)

You will need a large casserole with a lid.

1 BLOCK FIRM TOFU, PRESSED, PATTED DRY & CUT INTO 30 mm
CUBES

4 CUPS HOT WATER OR STOCK WITH 3 T WHITE (SHIRO) MISO
DISSOLVED IN IT

Dissolve into stock 1 t SAFFRON THREADS and simmer slowly in a saucepan.

¼ CUP HIJIKI SEAWEED, SOAKED 30 MINUTES IN WATER (SEE
INGREDIENTS & TECHNIQUES)

3 T OLIVE OIL

3 CLOVES GARLIC COARSELY CHOPPED

2 CUPS LONG-GRAIN BROWN RICE (WASHED AND DRAINED)

1 CUP FRESH GREEN PEAS

2 SWEET RED CAPSICUMS SLICED INTO LONG STRIPS

Preheat oven to 280°C (350°F), heat the olive oil in a large saucepan and sauté the garlic for 1 minute. Add the rice and sauté 2-3 minutes over medium-high heat. Add the remaining ingredients. Stir and cook 2-3 minutes. Place in a casserole. Cover and bake for 15 minutes. Meanwhile mix thoroughly:

ONE BLOCK TEMPEH

2 CLOVES GARLIC FINELY CHOPPED

1 t PREPARED CHILLI (SEE *INGREDIENTS & TECHNIQUES*) OR
TAO-TJO (TAUCHO) SAMBAL

2 T SHOYU

2 T WATER

With an apple-corer, deftly cut cylinders from block of tempeh. Crumble the rest and mix all with the garlic-chilli mix. Roll cylinders in wholemeal flour and deep fry until golden. Sprinkle tempeh 'sausage' on top, and the crumblings. Bake another 10 minutes. Serve with a red and green capsicum salad and blanched vegetables.

TOFU IN BLACK BEAN SAUCE

500 g TOFU

3 T PEANUT OR LIGHT SESAME OIL

5 CLOVES GARLIC FINELY CHOPPED

1 T FINELY GRATED GINGER

3 ONIONS COARSELY CHOPPED

3 T CHINESE BLACK BEANS (SEE *GLOSSARY*)

WATER

2 T MALTOSE OR RICE MALT OR 2 t HONEY

3 T KUZU DISSOLVED IN 1/3 CUP WATER

2 T TAMARI

2 t DARK SESAME OIL

In a wok, heat oil 'til nearly smoking. Add garlic and ginger. Stir fry rapidly over high heat for 1 minute. Add black beans, stir 30 seconds, add onions, and stir 2-3 minutes. Add tamari, maltose and 2 cups water. Bring to boil. Add kuzu. Stir with a wooden spoon 'til it thickens and becomes clear and shiny. Stir in tofu and dark sesame oil and simmer 2-3 minutes.

Serve on rice with blanched Chinese greens.

PASTA WITH TEMPEH BALLS (SERVES 6)

PASTA:

300 g DRY WEIGHT OF TUBES, SHELLS, SPIRALS OR SPAGHETTI,
  BUCATINI, TAGLIATELLE

Cook according to packet directions. Rinse under cold water and drain.

TOMATO SAUCE:

8 FRESH TOMATOES SKINNED

3 T OLIVE OIL

1 CLOVE GARLIC CRUSHED

2 BAY LEAVES

1 CUP WATER OR STOCK

1 t SALT

2 t OREGANO

3 T FRESH BASIL CHOPPED

(Greek oregano is best for this dish. It is sold in plastic bags with long flower heads of oregano — quite pungent. Somebody in the Italian community in Sydney grows it — I once got some fresh from a greengrocer in Kirribilli.)

Chop tomatoes very finely or process them. Heat olive oil and sauté garlic with bay leaves. Add tomatoes and salt and sauté over medium high heat for 5 minutes. Add stock and oregano, cover and simmer 30 minutes. Meanwhile:

½ BLOCK TEMPEH

2 CUPS COOKED BROWN RICE

1 CUP PEANUTS FINELY CHOPPED

2 T TAMARI

1 ONION FINELY DICED

Mash the tempeh with tamari. Add rice and peanuts. Add copious quantities of fresh black pepper. Knead all three together by hand until well mixed. Add diced onion and mix. Form into balls with wet hands and deep fry or bake.

The tomato sauce should be ready. Add the basil and the balls. Simmer 5 minutes. Ladle the mixture on to pasta. Garnish with coarsely-chopped, flat-leaf parsley.

BARLEY CORN TEMPEH (SERVES 6)

For this you need:

3 CUPS COOKED BARLEY

3 CUPS COOKED CORN KERNELS CUT OFF THE COB

¾ CUP COOKED HIJIKI SEAWEED (SEE *INGREDIENTS & TECHNIQUES*)

1 BLOCK DEEP-FRIED TEMPEH IN SMALL CUBES

3 T PEANUT, SAFFLOWER OR LIGHT SESAME OIL

3 CLOVES GARLIC FINELY CHOPPED

2 ONIONS FINELY DICED

1 RED CAPSICUM CUT IN SMALL SQUARES

1 CUP FRESH GREEN PEAS

3 T SHOYU

2 T MIRIN

½ CUP WATER OR STOCK

Heat oil in large saucepan, sauté garlic and onions 2-3 minutes. Add peas, capsicums and sauté 2 minutes. Add stock, mirin, shoyu. Cover and simmer over medium heat 5 minutes. Add remaining ingredients and heat through. Can be served pressed into a moistened cup and turned out as a dome.
Serve with stir-fried vegetables and salad.

HIJIKI OMELETTE (SERVES 6-8)

1 CUP COOKED HIJIKI CHOPPED (SEE *INGREDIENTS & TECHNIQUES*)

4-5 SHIITAKE MUSHROOMS (SOAKED AND SLICED)

3 T LIGHT SESAME OIL

1 ONION FINELY DICED

1 t GINGER FINELY GRATED

3 T MIRIN

2 T TAMARI

1 t RED SZECHWAN PEPPER (WELL-GROUND)

6 EGGS

3 CUPS SOYMILK

1 t DARK SESAME OIL

Heat oil in frying pan. Sauté onions and ginger 2-3 minutes. Add pepper, mirin, tamari. Cook 1 minute. Add shiitake and simmer. Whip eggs with soymilk and dark sesame oil, add hijiki. Pour into frypan and stir. Cook over low heat, lid off. Place under griller to complete. Alternatively, steam, or bake this mixture on a pastry base.
Serve with blanched vegetable salad.

COCONUT RICE, BAROBADUR TEMPEH, ACHAR (SERVES 8)

RICE:

4 CUPS LONG-GRAIN BROWN RICE

6 CUPS COCONUT MILK (FRESH OR CANNED)

1½ CUPS WATER

1 t SALT

Mix water and coconut milk and bring to a boil, stirring often to prevent it separating. When it boils, add salt and rice. Cover. Allow to re-boil, then simmer for a further 30 minutes.

ACHAR:

Ideally, make a lot of this and keep it refrigerated — it lasts 1-2 months and it gets better and better. You will need a selection of seasonal vegetables for this.

3 CUPS CAULIFLOWER IN SMALL FLOWERS

1 CUP GREEN BEANS SLICED

1 CUPS CUCUMBER IN SMALL CHUNKS

1 CUP CABBAGE SLICED

Blanch vegetables in boiling water and then rinse in cold water. Drain.

ACHAR SAUCE:

1-5 CHILLIES

3 ONIONS CHOPPED

5 CLOVES GARLIC CHOPPED

1 t (HEAPED) BLACHAN

250 g ROASTED PEANUTS

1 CUP RICE VINEGAR OR CIDER VINEGAR

3 T PALM SUGAR, MALTOSE OR 2 T HONEY

1 T TUMERIC POWDER

2 t SALT

½ CUP PEANUT OIL

Pound, process or pureé onions, garlic, blachan and chillies. Heat oil until almost smoking, sauté onion mix over medium heat for 10 minutes. Add remaining ingredients, vegetables and mix well. Simmer 5 minutes. Serve at room temperature or chilled.

BAROBADUR TEMPEH:

2 BLOCKS TEMPEH

MARINADE:

¼ CUP LOW-SALT SHOYU

4 T MIRIN OR SAKE

4 T WATER

6 CLOVES GARLIC FINELY CRUSHED

Mix all marinade ingredients. Add tempeh. Marinate at least ½ hour. Tempeh should be sliced straight through the block to make long ovals or fingers. Roll in wholemeal flour and deep fry briefly until golden.

ACCOMPANYING SALAD:

Arrange LETTUCE

BOILED POTATO

AGÉ TOFU PIECES

CUCUMBER CHUNKS

BEAN SPROUTS

SLICES OF EGG

. . . and a peanut dressing such as PEANUT SATAY SAUCE.

CAULIFLOWER TOFU

1 SMALL OR ½ LARGE CAULIFLOWER

This needs to be cut into medium-sized pieces and blanched. Bring enough water to the boil. Add 1 T caraway or dill seeds and sea salt. Plunge in cauliflower for 2-3 minutes, depending on thickness of stem. Drain. Arrange cauli on a baking tray. Squeeze a little lemon juice on it.

then, pureé

450 g TOFU

4 T WHITE MISO

1 CUP SOYMILK

2 T MIRIN

Pour this over the cauliflower and bake at 280°C (350°F) for 30 minutes or until it firms and gets browned.

Similarly:

MILLET, GREENS AND TOFU

Cook millet as for rice (see Ingredients & Techniques). Blanch greens (spinach), Chinese cabbage, silverbeet, broccoli, kale, big parsley, etc.) as for previous recipe.

PUREÉ:

450 g TOFU

2 T TAHINA

2 T LEMON JUICE

2 T WHITE MISO

PAPRIKA

Cover the bottom of a casserole with millet, spread a layer of the greens on it and pour the tofu on the lot. Sprinkle with paprika and bake as for previous recipe.

TOFU POLENTA

POLENTA:

1 ½ CUPS POLENTA OR CORNMEAL

5 CUPS WATER

1 t SALT

Dissolve salt in water and bring to a boil. Add polenta in a stream, stirring constantly with a whisk, until it thickens. Turn heat to low, cover and simmer 10 minutes, occasionally stirring with a wooden spoon. Pour into a rinsed dish or tray and leave to set two hours.

TOFU:

300 g TOFU

3 T WHITE MISO

1 T LEMON JUICE

½ CUP TOASTED WALNUTS

Use good walnuts and toast until oily but not browned. Pureé all of the ingredients with a little water or stock if necessary.

EITHER:

Pour tofu mix on the polenta and bake 30 minutes at 280°C (350°F). Cut into squares and serve.

OR:
Pour tomato sauce from a previous recipe (Pasta with Tempeh Balls) between polenta and tofu. You will have to carefully spoon tofu on top so as not to mix them. Bake as above.

TOFU FRIED RICE (SERVES 6)

6 T LIGHT SESAME OIL

4 GARLIC CLOVES CRUSHED

1 ONION FINELY CHOPPED

2 t GINGER FINELY GRATED

1 SMALL CARROT CUT JULIENNE (MATCHSTICKS)

1 SMALL RED CAPSICUM

1 CELERY STALK SLICED DIAGONALLY

A FEW CELERY LEAVES TORN UP

½ CUP ROASTED CASHEW NUTS

4 T SHOYU

5 CUPS PRE-COOKED LONG-GRAIN BROWN RICE

10 AGÉ TOFU CUBES SLICED

¾ CUP COOKED HIJIKI SEAWEED (SEE *INGREDIENTS & TECHNIQUES*)

2 T DARK SESAME OIL

In a wok, heat oil until almost smoky. Add garlic, onions and ginger. Stir fry over high heat for 2 minutes. Add carrot, capsicum and celery. Stir fry for 2 minutes, put top on wok, turn down heat and let the vegetables 'sweat' 2-3 minutes. Add rice and remaining ingredients except dark sesame oil. Stir fry until heated through. Add more shoyu if you like. Thoroughly mix in dark sesame oil. Serve.

BAKED VEGETABLES WITH TOFU AND MUSHROOMS

In a sound pottery or enamel container, bake matchbox-sized pieces of pumpkin, slices of onion, carrot, celery and daikon radish. Mix them with 2 T shoyu and 1 T mirin before sealing the lid and baking 1-1 ½ hours, depending on the oven. Alternatively, steam the vegetables.

400 g WHOLE, SMALL, BUTTON MUSHROOMS

4 T VIRGIN OLIVE OIL

5 CLOVES GARLIC CRUSHED

1 t DRIED BASIL

2 T SHOYU

2 T KUZU DISSOLVED IN 4 T WATER

½ CUP WATER OR STOCK

1 t SEA SALT

1 BLOCK TOFU CUT IN LARGE DICES

Wash mushrooms and drain. Pat dry. Heat olive oil, sauté garlic briefly. Add mushrooms, keeping heat high, with shoyu, salt and basil. Stir, replace lid and simmer mushrooms rapidly for 5 minutes so they lose lots of juice. Drop tofu briefly into boiling water, remove and rinse under cold water. This will give it a firmer texture.

Strain mushrooms. Put stock back on the stove and bring to a boil. Add the kuzu while stirring and stir until it thickens and gleams. Add the mushrooms and tofu and simmer 2-3 minutes. Pour over baked vegetables and mix gingerly so as to not mash them. Serve with rice, noodles or bread and salad. Good winter fare!

TOMATO TEMPEH (SERVES 5)

A clear tomato flavour with crunchy tempeh.

1½ BLOCKS TEMPEH

Deep fry or bake 200 mm cubes until golden and crisp.

8 FRESH RIPE TOMATOES

Plunge in boiling water and remove skins. Squeeze seeds into strainer and reserve liquid. Discard seeds. Chop tomato very fine and slowly bring to a boil with 1 T finely grated ginger and reserved liquid. Simmer 2-3 minutes. Add tempeh and salt to taste. Serve over rice or noodles.

NASI GORENG (SERVES 5)

5 CUPS OF DAY-OLD, LONG-GRAIN BROWN RICE

OR THAI JASMINE RICE
(IT MUST BE YESTERDAY'S OR THE DISH WON'T WORK)

4 T PEANUT OIL

1-3 RED CHILLIES SEEDED

2 CLOVES GARLIC

1 MEDIUM ONION

1 t BLACHAN

2 T TAMARI

OMELETTE STRIPS (SEE BELOW)

7 CUBES AGÉ TOFU **OR**

1 CUP SMALL TEMPEH CUBES

Pound, pureé or process the chillies, garlic, onion and blachan to paste. Heat oil and sauté the pureé 1 minute over high heat, then 5 minutes over low heat. Add agé or tempeh. Sauté 2 minutes. Add rice and tofu. Stir fry to break up rice. Add tamari and serve.

TO MAKE OMELETTE STRIPS:

Thoroughly mix 1 egg with ½ cup water or stock. Heat the wok fiercely and splash in this mixture. It will cook on contact. Tip out and break or slice to decorate Nasi Goreng.

Accompany Nasi Goreng with vegetables cooked in 50/50 water and coconut milk. Simply bring to a boil, add cabbage, carrot, green beans, cauliflower, broccoli and simmer 30 minutes. Salt to taste.

TOFU DAHL (SERVES 6-7)

Use skinned mung beans — moong dahl — for this if possible, or urid dahl or red lentils.

400 g SPLIT, HULLED MUNG BEANS
2 LITRES WATER
1 t TURMERIC POWDER

Wash the beans well and drain. Add to cold water with tumeric and bring to a boil. Turn down to a simmer. Skim scum and discard. Cook 30 mins.

3 T GHEE OR MUSTARD OIL, SESAME OR PEANUT OIL
1 t MUSTARD SEEDS
2 ONIONS
3 CLOVES GARLIC
1 t TURMERIC
1 t GARAM MASALA
2 T CORIANDER POWDER
1 T CUMIN POWDER
1 t CAYENNE OR CHILLI POWDER
SALT TO TASTE
1 BLOCK TOFU

Pound, pureé or process onions and garlic to a paste. Heat ghee and add mustard seeds. When the seeds pop, add onion and garlic to spices. Sauté over medium heat 5 minutes. Add tofu and mix well. As soon as the dahl is cooked, add to the spice-tofu mix and simmer 2-3 minutes. Salt to taste. If you can buy it, this dish is improved by the addition of ½ t asafoetidia (hing) with the spices. Hing reduces the flatulence factor in bean dishes.

TOMATOFU

Use the tomato sauce for the PASTA AND TEMPEH BALLS recipe, but add an extra cup of water or stock in which you have dissolved 1 level tablespoon of dark miso.

Use 1 block 450-500 g tofu cut into large cubes. Parboil to give the tofu a firmer texture. Add tofu to the sauce and simmer 2-3 minutes. Serve over cooked millet with fresh sweetcorn.

MADRAS TEMPEH

4 T PEANUT OIL

3 CLOVES GARLIC

1 t GINGER FINELY CHOPPED

2 ONIONS

1-5 CHILLIES

2 CINNAMON STICKS

2 t TURMERIC

2 PINCHES OF CURRY LEAVES (SEE *GLOSSARY*)

400 ml COCONUT MILK

1 t SEA SALT

4 CUPS SWEET POTATO IN MEDIUM LARGE CHUNKS

MADRAS CURRY POWDER

1 BLOCK TEMPEH IN SMALL SLICES OR DICES

Pound, pureé or process onions, garlic, ginger and chillies. Heat oil and fry onion paste over medium-high heat for 2-3 minutes. Add turmeric and curry and turn down heat. Cook for 2 minutes. Increase heat and add sweet potato and stir it in with the paste for 1 minute. Add tempeh, salt, coconut milk and cinnamon, plus 1 cup water. Bring to a boil while stirring slowly, and then simmer 30 minutes. Serve over rice with cucumber slices.

Better the next day!

SATAY TEMPEH (SERVES 6)

Unless you have a charcoal grill which can be fanned to white heat, so searing the food on sticks to earn the title 'satay' – grilling or deep frying will have to do.

Cut one block of tempeh into 20 mm cubes. Thread ⅓ of the skewer with cubes. Either grill, deep fry or char-grill the tempeh so it is slightly scorched. Take out the satay sauce you stashed in the refrigerator and bring it up to just above room temperature on a low heat. Dip the skewers in sauce and eat! Accompany with rice, raw and blanched vegetables. Manage to rest the skewer on the rice so the sauce seeps through it. Tofu works well in this recipe instead of tempeh.

SAUCE:

3 T PEANUT OIL

400 ml COCONUT MILK

1 CUP PEANUTS (ROASTED OR BOILED)

1-5 CHILLIES

1 ONION

5 CLOVES GARLIC

2 t CUMIN POWDER

2 t CORIANDER POWDER

1 t LAOS

2 T TAMARI

1 T GULA (PALM SUGAR)

1 T CURRY POWDER (MALAYSIAN)

1 T TAMARIND WATER (OR LEMON JUICE)

1 t SEA SALT (OPTIONAL)

Pound, pureé or process chillies, garlic, onion. Heat oil until smoking and saute paste 2 mins. Lower heat, saute 5 mins. Pureé peanuts with 2 cups water. Add to onion paste with remaining ingredients except coconut milk. Mix well and cook 2-3 mins. Add coconut milk and stir slowly for 10 minutes. Keeps a long time and improves considerably with age. May I suggest making double the recipe?

STIR-FRIED TOFU

4 T LIGHT SESAME OIL (OR 4 T WATER FOR AN OIL-LESS DISH)

3 CLOVES GARLIC POUNDED OR 1 t FINELY GRATED
FRESH GINGER

1 MEDIUM CARROT IN MATCHSTICKS

1 CUP CHOPPED CABBAGE

100 g BEAN SPROUTS

1 MEDIUM-SMALL MUSTARD CABBAGE SLICED

2 T SHOYU

1 t DARK SESAME OIL

1 T KUZU DISSOLVED IN ¼ CUP WATER

1 BLOCK TOFU IN 10 mm THICK RECTANGLES

Wash mustard cabbage thoroughly. In a wok, heat oil, sauté garlic briefly, and add carrots and cabbage. Stir fry over high heat 2 minutes. Add mustard cabbage and ¼ cup water. Clap the lid on. Let it sizzle for 1 minute, add shoyu and tofu and stir in. Cover and simmer over medium heat for 2-3 minutes. With your wok instrument, shovel the vegetables up to the side of the wok, leaving the stock in the bottom. Turn heat to low and add the dissolved kuzu, stirring it thoroughly. As soon as it thickens and starts to clear, stir the vegetables in it so as to coat them with glaze. Simmer for 1 minute and serve over noodles or rice.

VEGETABLES WITH WHITE SOY SAUCE

Steam, boil or bake a selection of vegetables such as pumpkin, potato, carrot, sweet potato, onion, broccoli, cauliflower, etc. Serve them with the following sauce over rice, millet, noodles or spaghetti. Alternatively, cover them with the sauce and serve as part of a buffet.

2½ CUPS SOYMILK

3 T WHITE MISO

2 T ALMOND BUTTER (OR PASTE, PREF ROASTED)

3 T KUZU DISSOLVED IN 6 T WATER AND A LITTLE OF THE SOYMILK

Cream together almond butter and miso. Bring soymilk to high heat, add kuzu, stirring rapidly until it clears and thickens. Turn down heat and simmer for 5 minutes. Remove a few tablespoons of sauce and cream the almond and miso with it. Incorporate back into sauce.

TEMPEH MEE GORENG (SERVES 5)

'Mee' are spaghetti-size Chinese wheat noodles. They can be bought fresh, refrigerated, from a Chinese food store. Douse them with hot water if they are oily. Alternatively, replace them with Japanese brown rice noodles — genmai udon or udon. Spaghetti can be substituted if necessary.

3 T PEANUT OIL

2 ONIONS

3 CLOVES GARLIC

1 t BLACHAN

1 CUP CABBAGE IN THICK STRIPS

8 CUBES AGÉ TOFU SLICED

½ BLOCK TEMPEH IN SMALL THIN RECTANGLES

3 T SHOYU

BLACK PEPPER

1-2 T PENANG OR MALAYSIAN CURRY POWDER

1 t LAOS POWDER

1 T RICE MALT, MALTOSE OR PALM SUGAR

1 PKT BROWN RICE UDON PRE-COOKED AND RINSED (OR FRESH MEE)

1 CUP MUNG BEAN SPROUTS

LEMON WEDGES TO GARNISH

Pound, process or pureé the onion, garlic, blachan and, if you want it hot, 1-3 chillies. Heat oil in a wok and stir fry the paste for 2-3 minutes. Turn down heat and let mixture simmer 1 minute. Raise heat, add spices, and rice malt. Stir. Add cabbage, tempeh and tofu. Stir fry 3 minutes. Use a little water or stock if it becomes too dry. Add the noodles and shoyu. Stir together and mix in the noodles and bean sprouts gently. Heat 1 minute more and serve, garnished with lemon to squeeze over individual serves.

TABRIZ TEMPEH (SERVES 5)

1 BLOCK TEMPEH

2 MEDIUM TOMATOES (PEELED AND SEEDED)

2 RED CAPSICUM VERY FINELY DICED

1 ONION FINELY DICED

1 CLOVE GARLIC CRUSHED

2 T OLIVE OIL

½ t OREGANO DRIED

½ t CUMIN POWDER

1 t CORIANDER POWDER

1 T TAMARI

1 T RED WINE

SEA SALT

Break up the tempeh so it is almost back to the bean. Chop the tomato very finely. Heat oil and sauté garlic and onion for 1 minute. Add tempeh and sauté for 5 minutes, stirring until tempeh browns. Add wine and tamari, tomatoes, capsicum, spices and salt. Sauté 10 minutes over medium heat. Serve over rice, with flat bread or with a millet pancake — simply substitute ½ millet flour in your favourite savoury pancake recipe.

THAI PUMPKIN AND TOFU

⅓ MEDIUM-SIZED SWEET PUMPKIN CUT TO MATCHBOX SIZE

1 LEMONGRASS STALK OR ONE HANDFUL DRIED LEMONGRASS

   OR 1 T POWDER

2 CUPS COCONUT MILK

2 TABLESPOONS NAM PLA

1 t LAOS POWDER

450 g TOFU DICED LARGE

1 T MINT CHOPPED

1 HANDFUL FRESH CORIANDER TORN

1 T LIME JUICE

Chop and pound lemongrass. Mix mint and coriander with juice. Place Laos, lemongrass, pumpkin, nam pla and milk in a saucepan. Slowly bring to a boil while stirring. Simmer until pumpkin is just soft. Add tofu, stir in and simmer until tofu is hot. Turn on to a plate, sprinkle with mint and coriander and juice. Soymilk can be substituted for coconut milk. Serve with Thai jasmine rice and blanched greens.

TOFU 'PIZZA'

Either use a sourdough or yeasted base. The sourdough base tastes better, but you need a starter for it. If you have the leaven, or starter, simply make your normal bread dough, but roll it out like a thick pastry and place in a baking tray to rise. Brush with olive oil. Pre-bake for 5 minutes before spreading on topping. Refer to the NATURAL TUCKER BREAD BOOK *for details on sourdough.*

These pizzas also work on a simple flat-bread base.

Otherwise:

500 g 80% FLOUR (OR ½ UNBLEACHED WHITE, ½ WHOLEMEAL)

10 g YEAST (FRESH)

10 g SEA SALT

2 T OLIVE OIL

1 ⅓ CUPS WATER (WARM)

In a large bowl, rub oil into flour. Cream yeast with a little water and mix with flour. Dissolve salt in remaining water and mix in. Work it with a wooden spoon and then knead with your hands for 5 minutes, until it's smooth and elastic. Place in a floured bowl, enclose in a plastic bag and — go do the shopping. Three hours' fermentation will produce a 'ripe' dough. Punch it down and re-knead for 2-3 minutes. Roll out to fit your pizza tray and brush with olive oil. Leave for 30 minutes to rise. Pre-bake 2-3 minutes at 200°C (400°F). Use any leftover dough to make rolls or mini pizzas for lunches.

SAUCE:

See PASTA AND TEMPEH BALLS, *or simply use a good tomato paste or spread with miso sauce.*

TOFU TOPPING:

250 g TOFU PRESSED

3 T WHITE MISO

1 T OREGANO

BLACK PEPPER

OLIVES CHOPPED, OLIVE STOCK

SLICED MUSHROOMS

CRUSHED WALNUTS

Pureé, process or pound tofu and miso. Thin with a litte olive stock (2-3 T). Spread tomato sauce, then tofu on pre-baked base. Grind copious black pepper on top, sprinkle oregano, sliced olives, mushrooms and crushed walnuts. Pat them into the tofu slightly. Bake at 205°C (425°F) for 10-15 minutes, depending on base thickness.

OR:

Crumble the tofu and stir roughly with all ingredients except miso. Substitute salt or olive paste if you wish. Sprinkle on and compress it slightly. Bake as above.

TEMPEH 'PIZZA'

Follow the directions for TOFU 'PIZZA' *right up until you have spread the tomato sauce on the base.*

TEMPEH:

'Shave' the tempeh to resemble large, thin pencil shavings, or cut out shapes with and apple corer to resemble sausage rounds. Crumble the remainder. Marinate for at least ½ an hour in . . .

4 T TAMARI
2 T TOMATO PASTE
3 CLOVES GARLIC FINELY CRUSHED
1 t DARK SESAME OIL
1 t GINGER FINELY GRATED
1 t CIDER VINEGAR
1 t HONEY
A LITTLE CHILLI PASTE IF DESIRED

Drain well and sprinkle on tomato sauce with sliced mushrooms and sliced olives. Sprinkle oregano and fresh black pepper. Bake as for TOFU 'PIZZA'. *Finish it off under the grill if it isn't crisp enough. If you like, mix some mozzarella and parmigiano with the tempeh.*

An alternative to tomato sauce or paste is to slice a few rounds of tomato under this topping, pre-grill briefly to 'dry' them slightly.

STIR-FRIED YUBA

Soak yuba sticks for 2 hours with a little bi-carb of soda to minimise beany flavour. Rinse, soak for 3 hours more or overnight. Drain well. Put in a tea towel and gently wring dry. Chop or tear into 70 mm lengths.

3 T LIGHT SESAME OIL

1 T GINGER FINELY GRATED

1 MEDIUM DAIKON CHOPPED INTO MATCHSTICKS

1 CUP CHOPPED WATERCRESS (80 mm LENGTHS)

100 g BEAN SPROUTS

1 SMALL HEAD BROCCOLI, FLOWERS SEPARATED, STALK
    SLICED THINLY

2 T SHOYU

1 t DARK SESAME OIL

In a wok, heat oil and briefly sauté ginger. Add yuba and fry hard for 2-3 minutes. Add remaining ingredients except sesame oil. Fry 2-3 minutes, leave lid on for 1 more minute. Stir. Add sesame oil, stir it in and serve on rice or noodles.

TOFU TARTS OR FLAN

PASTRY:

4 CUPS WHOLEMEAL FLOUR

½ CUP OIL (INCREASE THIS TO ¾ CUP FOR 'SHORTER' PASTRY
    AND REDUCE WATER)

PINCH SEA SALT

1⅓-1½ CUPS WATER (DEPENDING ON FLOUR)

Add salt to flour, rub in the oil until well mixed. Add water, firstly 1 cup then the remainder as needed. Mix in with a wooden spoon until it clings together. Then mix with hands to form a dough. This must not be hard, but pliable, so add more water if necessary (wholemeals vary a lot in their capacity to absorb water). Knead it — yes, for 1 minute. This can be used immediately, but is also good kept sealed for 3 days, after which it becomes sourdough and much lighter, with a pleasant tang. Use either tart tins with removable bases or one large flan tray. Oil tins, roll out the pastry and line your tin(s) with it.

FILLING:

These are numerous but I will describe a walnut-spinach mix.

Wash thoroughly 1 bunch spinach. Cut off pink roots. Bring 3 cups of water to the boil and add spinach. Reboil and simmer for 5 minutes. Drain and plunge spinach into cold water. When it's cool, drain well and chop it up.

2 ONIONS CHOPPED

3 CLOVES GARLIC CRUSHED

2 T OLIVE OIL

1 T CORIANDER POWDER

3 T SHOYU

2 t OREGANO

FRESH-GRATED NUTMEG — A FAINT SPRINKLE

1 CUP PRE-COOKED RICE OR MILLET

1 CUP WALNUTS CHOPPED

COPIOUS FRESHLY GROUND BLACK PEPPER

Heat oil, fry garlic and onions for 2 minutes. Add remaining ingredients except walnuts and spinach. Fry 2-3 mins. Add spinach and walnuts, mix well. Drain and reserve liquid.

TOFU:

250 g TOFU

1 T WHITE MISO

1 T MIRIN

2-3 T DRAININGS FROM SAUTÉ

Pound, process or pureé all above ingredients. Place spinach filling in tart. Spoon on tofu mix and smooth with spatula. Bake at 200°C (400°F) for 15 minutes or until the tofu sets and is browned. Finish under griller if necessary.

TEMPEH ROLLS (MAKES 8-10 ROLLS)

1 BLOCK TEMPEH

1 PACK PHYLO PASTRY (WHOLEMEAL IF POSSIBLE)

BEAN SPROUTS

LETTUCE THINLY SLICED

SATAY SAUCE

CUCUMBER FINGERS

Grill, bake or deep fry sticks of tempeh until browned. On a square of phylo pastry, arrange enough vegetables for one roll. These should run in the centre, facing corners so the roll can be folded from one corner to the next and the other corners folded in, spring roll style.

Arrange tempeh on vegetables, spoon on satay sauce. Roll up and stand the roll on its join to ensure sealing. Brush with peanut or light sesame oil. Bake at 205°C (425°F) for 10-15 minutes.

TOFU AND TEMPEH BROCHETTE OR KEBABS

1 BLOCK TOFU IN LARGE DICES

1 BLOCK TEMPEH IN LARGE DICES

2 ONIONS IN QUARTERS, LAYERS SEPARATED

1 EACH RED AND GREEN CAPSICUM CUT IN SQUARES

Arrange tofu and tempeh on the skewer, alternating and spacing with onion and capsicum.

MARINADE:

½ CUP SHOYU

1 T GARLIC CRUSHED

1 T GINGER FINELY GRATED

2 t DARK SESAME OIL

2 t HONEY OR RICE MALT (2 T)

2 T TOMATO PASTE

1 T RICE VINEGAR

1 T FRESHLY GROUND CORIANDER

MUCH FRESHLY GROUND BLACK PEPPER

½ CUP WATER

Bring this slowly up to high heat and pour over skewers in a flat dish. Marinate for 1 hour and grill until well browned, turning occasionally.

TOFU 'MOUSSELINE' WITH SEAFOOD

500 g FRESH GREEN PRAWNS, SHELLED AND DEVEINED

  (OR FISH, SCALLOPS)

600 ml TOFU PUREÉ

1 EGG

2 t SPRING ONION VERY FINELY CHOPPED

½ t DARK SESAME OIL

Pound, pureé or process prawns to a paste. Mix egg and stir in to tofu with spring onion. Mix well. Add paste and oil and combine all ingredients thoroughly (re-process, even). Use either cups or stainless steel moulds if you have them. Lightly oil moulds and almost fill them. Place in a hot-water bath, cover the lot with foil or a lid and bake at 130°C (350°F) for about 30 minutes or until set. Turn out of moulds on to a plate for service.

SAUCE:

1 STICK KOMBU SEAWEED SOAKING IN

  450 ml WATER

  3 SHIITAKE MUSHROOMS, PRE-SOAKED

  2 t MIRIN

  2 T WHITE MISO

  1 t GINGER FINELY GRATED

  1 T KUZU DISSOLVED IN 3 T WATER

Add shiitake to water. Bring to the boil and simmer for 10 minutes. Strain and keep stock. Use kombu and mushrooms elsewhere. Add mirin, miso and ginger. Mix well. Over medium heat, add the dissolved kuzu, stir till clear and thick. Simmer over very low heat for 5 minutes.

WATERCRESS

HIJIKI SEAWEED

2 T RICE VINEGAR

2 T SHOYU

Soak 1 cup of long strands of hijiki seaweed for ½ hour (see INGREDIENTS & TECHNIQUES). Drain and douse with rice vinegar and shoyu mixed in equal proportions. Clean watercress and break into attractive lengths. Arrange hijiki and watercress around the base of the mousseline and ladle some sauce on to each.

ALTERNATIVELY:

250 g GREEN PRAWNS

250 g FISH FILLET

600 ml TOFU PUREÉ

2 EGGS

1 T NAM PLA

1 T LIME JUICE

2 T CHOPPED BASIL LEAVES

1 GREEN CHILLI SEEDED

Pureé basil, chilli, lime juice and nam pla. Proceed as above.

SAUCE:

400 ml COCONUT MILK

2 t LAOS POWDER

OR 2 T POWDER OR 1 STICK LEMONGRASS CHOPPED AND POUNDED

SALT TO TASTE

Slowly bring all ingredients to the boil while stirring. Simmer 15 minutes. Strain and ladle over mousseline with hijiki, but replace watercress with coriander.

A CHILLED VERSION WITH NO EGG

2 T AGAR FLAKES

1 CUP WATER

Soak the flakes with water in a small saucepan for 5 minutes. Very slowly heat while stirring until all agar is dissolved. Immediately mix with:

500 ml TOFU

2 T WHITE MISO

¾ CUP TOASTED WALNUTS

. . . and pureé or process thoroughly. Pour into rinsed moulds to set in refrigerator. This should take 1 hour.

SAUCE:

3 T DIJON MUSTARD	
3 T NATTO MISO	
1 T TAHINA	
6 T WATER	

Mix thoroughly. Serve on a bed of iced soba (buckwheat) noodles, garnished with watercress, on a hot day.

TOFU BURGER

This is an arrangement much in the same fashion as a hamburger. Tofu is best grilled for this (see GRILLED TOFU). Use the tofu mayo sauce as dressing. Fill with tomato, alfalfa sprouts, lettuce, cucumber, vinegared onion rings, etc.

TEMPEH BURGER

As above, but deep fry or grill the tempeh. Use satay sauce.

SALADS & DRESSINGS

Salads need little explanation and so I only include a few recipes — the dressings are more significant. Also refer to APPETIZERS & DIPS.

DRESSINGS

WHITE MISO:

1/2 CUP WHITE MISO

2 T BROWN RICE VINEGAR OR LEMON JUICE

1 T TAHINA

Thin to desired consistency with water or apple juice.

1/2 CUP WHITE MISO

2 T PREPARED MUSTARD

1 T HAZELNUT OR ALMOND BUTTER

Mix well.

1/2 CUP WHITE MISO

2 T RICE VINEGAR

1 t TAHINA

2 t MUSTARD

Mix and thin with water or apple juice.

FOR CHICKPEA SALAD:

1 T TAHINA

3 T LEMON JUICE

1 t SHOYU OR MISO

Mix well.

TOFU MAYO (2)

250 g TOFU

3 DILL PICKLES (GHERKINS)

1 T LEMON JUICE

1 T MUSTARD

1 T RICE MALT OR 2 t MAPLE SYRUP

A LITTLE WATER

Pureé all ingredients.

HIJIKI AND TOFU SALAD

1 BUNCH RADISHES

HANDFUL STICK BEANS

COS LETTUCE

HANDFUL MUNG BEAN SPROUTS

1 CUP COOKED HIJIKI (SEE *INGREDIENTS AND TECHNIQUES*)

2 RED SALAD ONIONS

RICE OR CIDER VINEGAR

450 g TOFU DICED

Parboil the tofu and cool. Blanch the beans and cool. Soak onion rings in rice or cider vinegar with a pinch of salt for 10 mins. Arrange cos leaves in a glass bowl. Mix all ingredients with tofu mayo and fill the bowl or serve mayo separately.

WAKAME

1 CUP WAKAME SOAKED FOR 15 MINUTES

2 CARROTS JULIENNED (MATCHSTICKS)

1 CUP CABBAGE SHREDDED

1 RED CAPSICUM CUT IN STRIPS

1 CUP DAIKON JULIENNED AND SOAKED IN COLD WATER
    FOR 30 MINUTES

1 CUP MUNG BEANS OR AZUKI SPROUTS

Combine all ingredients and use a white miso dressing.

CUCUMBER AND SESAME

3 CUPS CUCUMBER CUT IN QUARTERS LENGTHWISE. SLICE
    THROUGH IN 10 cm SECTIONS

1 CUP WAKAME SOAKED FOR 15 MINUTES

2 T RICE VINEGAR

3 T TOASTED SESAME SEEDS

Mix vinegar with wakame and add remaining ingredients. Omit the cucumber for a different salad and add 2 t shoyu.

TEMPEH SALAD

Parboil, steam or (preferably) deep fry 1 block cubed tempeh. Put in 'Devilled' Marinade for 1 hour. Combine with cucumber, diced potato, hard-boiled egg, bean sprouts and blanched cabbage. Serve on lettuce.

FENNEL SALAD

½ BULB FENNEL THINLY SLICED

2 TOMATOES

½ CUP OLIVES

ENDIVE BROKEN IN PIECES

450 g TOFU CUBES

1 SALAD ONION SLICED IN RINGS AND SOUSED WITH VINEGAR

FLAT-LEAF PARSLEY COARSLEY CHOPPED

2 T OLIVE OIL

SALT

Marinate tofu for 30 minutes with onion and some of its pickling vinegar. Mix 1 T of the vinegar with olive oil and salt and whisk together. Drain tofu and mix with other ingredients. Alternatively, marinate tofu in olive brine.

THAI SALAD

2 CUPS PEANUTS, ROASTED AND COARSELY CHOPPED

½ SHREDDED COS LETTUCE

250 g BEAN SPROUTS

3 T FRESH BASIL CHOPPED COARSELY

3 T FRESH MINT CHOPPED COARSELY

3 T FRESH CORIANDER CHOPPED COARSELY

3 T NAM PLA

2 T PALM SUGAR

JUICE OF 2 LIMES

450 g TOFU CUBED OR TEMPEH DEEP FRIED OR BAKED

If using tofu, mix nam pla, palm sugar and lime juice. Marinate tofu cubes for 15 minutes. Mix with remaining ingredients as above and combine with vegetables and herbs. The tempeh doesn't need marinating — it will lose its crisp texture if marinated.

DESSERTS

TOFU 'CHEESE' CAKE

TOPPING:

500 g TOFU

2 t PURE VANILLA ESSENCE

2 FREE-RANGE EGGS

1 LEMON — JUICE AND RIND

½ CUP MAPLE SYRUP

1½ T ARROWROOT

CRUST:

2 CUPS 80% WHOLEMEAL FLOUR

½ CUP DESICCATED COCONUT OR ROLLED OATS

¼ CUP OIL

2 T RICE MALT OR BARLEY MALT

½ CUP APPLE JUICE

½ t SALT (OPTIONAL)

Mix flour, oats and salt. Rub in oil. Dissolve malt in apple juice. Add to flour mixture and mix thoroughly. Spread or roll on the bottom of the pie tin as base. Pre-bake for 5 minutes at 180°C (350°F). Put all topping ingredients in a blender or processor and pureé thoroughly. Pour on to base. Bake for 30-40 minutes at the same temperature. Alternatively, don't make the base, but bake the 'cake' in an oiled fluted metal or glass shape. Cook and simply turn out and decorate with strawberries, raspberries and blueberries.

RICH SOY ICE-CREAM

8 EGGS SEPARATED

5⅓ CUPS TOFU PURÉE

2 CUPS MAPLE SYRUP

4 t PURE VANILLA ESSENCE

Beat together the egg yolks and while beating, add 1½ cups tofu pureé and 2 cups maple syrup. Put in a double boiler and stir gently until this thickens. Chill for 30 minutes. Mix vanilla with remaining tofu pureé. Fold into custard. Beat egg whites until peaks form and fold into tofu mix. Freeze immediately.

This is the basic mix and begs for strawberries and other fruit. Make it carob or chocolate by whisking 4 T pure chocolate powder into the syrup before making the custard. Similarly, mix 1 cup mashed (NOT pureéd) banana into the custard and cook.

CHERRY RIPE

This has to be made in two stages. In the first, make the base and the cherry agar. When this starts to set, make the topping. Alternatively, omit the base.

BASE:

2 CUPS WHOLEMEAL FLOUR

⅓ CUP LIGHT SESAME OR SAFFLOWER OIL

1 CUP DESICCATED COCONUT

1 CUP ROLLED OATS

¾ CUP APPLE JUICE

PINCH SEA SALT

Proceed as for the base of TOFU 'CHEESECAKE' *but fully bake it for 20 minutes.*

CHERRY:

½ KILO CHERRIES (STONED!)

3 T AGAR FLAKES

½ CUP COCONUT THREADS

1 CUP COCONUT MILK OR SOYMILK

½ CUP RICE MALT OR MALTOSE

2 T PURE MAPLE SYRUP

2 CUPS APPLE JUICE

Over medium heat, cook agar with apple juice until it dissolves. Add maple syrup, malt, coconut and cherries, bring to near boil. Add coconut or soymilk (or a combination) and stir slowly for 2-3 minutes. Pour on base and allow to set.

TOPPING:

½ CUP CAROB POWDER OR PURE CHOCOLATE POWDER

2½ CUPS WATER

2 t PURE VANILLA

3 T RICE MALT OR MALTOSE

2 T KUZU (HEAPED) OR 3 T ARROWROOT

½ CUP SOYMILK

1 T TAHINA

In a saucepan, add carob or chocolate to 2 cups water and whisk until well mixed. Slowly bring to near boil and simmer very slowly for 20 minutes. Add malt and dissolve it. Mix remaining water and soymilk with vanilla and kuzu. Stir well until dissolved. Increase heat under carob syrup and stir in kuzu mix until it thickens. Simmer. Take out a few tablespoons and thoroughly whisk tahina with it. Add back to carob and mix well. Spread over the cherry mix and refrigerate until firm.

SOY-PIPELINE CUSTARD

2½ CUPS SOYMILK

2 CUPS PUREÉD TOFU

½ CUP PURE MAPLE SYRUP OR ⅓ CUP HONEY

1 CUP NATURAL SULTANAS

NUTMEG AND CINNAMON TO TASTE

5 T KUZU

2½ CUPS APPLE JUICE

½ CUP ALMONDS ROASTED AND SLICED

250 g PRE-COOKED WHOLEMEAL MACARONI — THE LONGEST
　　YOU CAN GET, PREFERABLY WITH FLUTES

Bring soymilk, tofu, maple syrup, sultanas and spices to high heat. Thoroughly dissolve kuzu in apple juice. Add to soymilk mix and stir constantly until thickened. Mix almond slivers through macaroni with a little cinnamon. Pour mixture over macaroni, cool and refrigerate. Cooked apples make a delicious base and cinnamon sprinkled on top perfects the custard.

MISO-OAT BARS

3 CUPS ROLLED OATS

2 CUPS WHOLEMEAL FLOUR

2 T BARLEY MISO

½ CUP LIGHT SESAME OIL

2 CUPS DATE PURÉE

½ CUP APPLE OR GRAPE CONCENTRATE

2 t CINNAMON

2 t LOTUS ROOT POWDER (IF AVAILABLE) OR JUST USE CINNAMON

1 T ORANGE ZEST

Combine rolled oats, flour, spices and rub in oil thoroughly. Mix remaining ingredients together thoroughly. Blend wet and dry. Smooth into an oiled biscuit tray. With a knife, cut into bars. Bake at 180°C (350°F) for 20-30 minutes.

SOY-SEMOLINA

1 CUP SEMOLINA (FINE)
2½ CUPS SOYMILK
1 CUP WATER
RIND OF 1 LEMON
1 CUP RAISINS
½ CUP ALMONDS TOASTED AND CHOPPED
2 T TAHINA
SEA SALT TO TASTE
1 T HONEY
1 t PURE VANILLA ESSENCE

Add semolina to water and soymilk mix. Slowly bring to the boil while stirring, add the remaining ingredients except tahina. Simmer for 10 minutes. Take out a cup of the mix and blend tahina thoroughly. Mix back in and stir well. Pour into a rinsed dish to cool. Refrigerate or, alternatively, serve hot. Fruit can be served on top of this as a sauce or attractively set in an agar jelly.

Sauce:

500 g APPLES
1½ T KUZU OR 3 T ARROWROOT
1¼ CUPS APPLE JUICE

Core and slice apples. Cook in 1 cup apple juice until slightly soft. Dissolve kuzu in ¼ cup of juice. Strain apples. Thicken the juice with kuzu and stir until it clears. Add apples, mix and pour over semolina. This will set and can be cut into squares.

Fruit Jelly:

5 CUPS WHITE GRAPE OR APPLE JUICE
1 LEMON, RIND AND JUICE
3 T AGAR FLAKES
1 T KUZU OR ARROWROOT
1 PUNNET STRAWBERRIES

Dissolve agar in 4½ cups grape juice. Slowly bring to boil, add lemon rind and simmer until agar dissolves. Increase heat. Thoroughly mix kuzu with remaining juice and lemon juice and stir in to agar mix. Stir well until slightly thick and clear. Pour a very little on top of the semolina. Arrange cut strawberries on this and allow this thin film to set. Refrigerate if necessary for 5 minutes. Pour remaining mix over strawberries and cool. Refrigerate before serving.

BLUEBERRY SUNSET

4 CUPS APPLE JUICE

5 T AGAR FLAKES

4 T RICE MALT OR MALTOSE

2 T (HEAPED) KUZU

2 CUPS SOYMILK

1 t PURE VANILLA ESSENCE

1 PUNNET BLUEBERRIES

Over medium heat, cook together apple juice, agar flakes and malt until agar dissolves. Add kuzu to soymilk with vanilla and dissolve well. Increase heat. Add kuzu mix to agar, stirring constantly. When it thickens and becomes shiny, add washed blueberries and stir until the colour comes out. Pour in a rinsed dish or mould to set. This will set in 1 ½ hours at room temperature and a little faster in the refrigerator.

SOY BRANDY SAUCE (FOR SAGO PLUM PUDDING)

2 CUPS SOYMILK

1 ½ T KUZU DISSOLVED IN 3 T WATER

1 t HONEY OR (PREFERABLY) 2 t MAPLE SYRUP

1 T HAZELNUT PURÉE (SEE BELOW)

2 T BRANDY

To make Hazelnut purée:

Roast hazels in the oven until fragrant and very slightly browned. Cool them but while still warm, purée in a vitamizer or food processor until they have the texture of peanut paste (butter). If necessary, add a little bland oil such as safflower. Bring soymilk to a near boil, add maple syrup, stir in and add kuzu. Stir constantly until thickened. Take out ½ cup and mix in hazelnut purée and brandy until well creamed. Return to heat and simmer very slowly for 1 minute. I can't really give a recipe for a sauce without one for what it goes on!

SAGO PLUM PUDDING

2 CUPS COOKED SAGO
500 g PRUNES CHOPPED
1 CUP STOUT (GUINNESS OR COOPERS)
1 EGG BEATEN
¼ t NUTMEG FRESHLY GRATED
2½ CUPS 80% WHOLEMEAL FLOUR
½ t BI-CARB SODA
RIND OF 1 ORANGE (ZEST ONLY)
4 T MAPLE SYRUP OR RICE MALT (LESS SWEET BUT VERY GOOD)
2 T SAFFLOWER OR LIGHT SESAME OIL
2 t PURE VANILLA ESSENCE

To cook sago:

There are a few methods, but only this one is suitable here. Bring 2 litres of water to a rolling boil and tip in 2½ cups sago, stirring to prevent clumps. Simmer rapidly until the sago is clear, stirring occasionally. Pour it through a close-grained sieve and rinse thoroughly with cold water. These clear beads are also the basis of a Buddhist fish roe imitation.

The pudding:

Mix flour and soda thoroughly with nutmeg. Add prunes. Whisk or blend together the maple syrup, oil, vanilla, egg and orange rind. Mix the stout in and pour into flour mixture. Add sago and mix thoroughly with a wooden spoon. Tip into an oiled and floured (with 100% wholemeal) pudding steamer, clamp the lid or foil on and boil in water for 1½ hours. Make sure the water doesn't cover the lid.

THE BUDDHIST FISH ROE IMITATION

1 CUP SAGO BEADS (ABOVE)
2 T NORI TSUKEDANI (SEE *GLOSSARY*)

Mix thoroughly and refrigerate sealed for 1 week.

RICOTTA CREAM

2 CUPS RICOTTA
1 CUP SOYMILK
1 T PURE VANILLA ESSENCE (OPTIONAL)

Pureé together and serve as a dessert cream for pies, crumbles, stewed fruit, etc.

TOFU 'MOUSSE' WITH STRAWBERRIES

600 ml TOFU PURÉE (PREFERABLY SILKEN TOFU)

1 EGG

3 T MAPLE SYRUP

1 T PURE VANILLA ESSENCE

Pureé all ingredients thoroughly and spoon into cups. Place in a hot water bath, cover the lot with foil and bake at 200ºC (400ºF) for 30 minutes or simmer on the stove for 30 minutes. Serve hot or cold.

Sauce:

1 PUNNET STRAWBERRIES WASHED (CUT LARGE ONES IN HALF)

½ CUP APPLE JUICE

3 T RICE MALT OR MALTOSE

Put strawberries in a saucepan with apple juice, cover and slowly bring to the boil. When boiling, add rice malt. Mix thoroughly, cool and refrigerate. It gets better!

PARFAIT

I rarely make these from anything but leftovers — and these are always the best. From the recipes in this section, we can concoct a great combination of leftovers. Use carefully collected remains from SOY SEMOLINA, RICH SOY ICE-CREAM, FRUIT JELLY (on Soy Semolina). Layer them in a parfait glass and ladle STRAWBERRY SAUCE (from TOFU MOUSSE) over the lot.

TAHU FAH

A specialty of street vendors throughout South-East Asia. It is offered with hot ginger tea and palm sugar in rural areas, and sometimes with sago, agar strips, nuts, candied fruit as well .

GINGER TEA:

2 T GINGER FINELY GRATED

3 CUPS WATER

4 T PALM SUGAR

Place ginger in a saucepan with cold water and bring to the boil. Simmer 10 minutes. Add palm sugar and dissolve. Strain. Ladle over cold tahu fah.

A SOYFOODS CHRISTMAS CAKE (SOURDOUGH)

(makes 2 large cakes)

500 g RAISINS

750 g SULTANAS

250 g CURRANTS

125 g DATES OR FIGS CHOPPED

250 g ALMONDS

500 g TOFU

2 CUPS SAFFLOWER OIL

1 CUP RICE MALT OR MALT EXTRACT

4 T BRANDY OR RUM

1 t VANILLA ESSENCE

1 kg 80% OR WHOLEMEAL FLOUR

½ t CINNAMON

½ t ALLSPICE

½ t NUTMEG

PINCH OF CLOVES

2 CUPS LEAVEN (SEE *BELOW*)

½ CUP BARLEY MISO

RIND AND JUICE OF 1 ORANGE

This is best made with leaven or sourdough. Refer to NATURAL TUCKER BREAD BOOK *for details. Otherwise, use 2 t bi-carb soda mixed with the flour and 1 cup of dark grape juice to replace the liquid quantity of the leaven. When mixed, bake covered with foil at 200ºC (400ºF) for 1 hour. Remove foil and bake for 45 minutes at 180ºC (350ºF).*

Otherwise:

Purée together tofu, oil, malt, brandy, vanilla, miso, rind and juice. Mix the flour and spices. Add wet to dry and mix thoroughly, including the leaven. Stir in dried fruit and nuts. Place in tins which should not be too deep. Cover with a damp cloth, put in a warm spot and leave to rise for 3 hours. Bake at 180ºC (350ºF) for 1 hour covered with foil, then for 1 hour at 150ºC (300ºF) uncovered.

APPLE TOFU CAKE

500 g WHOLEMEAL FLOUR

1 t BICARB SODA

¾ CUP TOFU PURÉE

1 CUP APPLE JUICE

1 CUP HONEY OR MAPLE SYRUP

1 T CINNAMON

¼ t NUTMEG FRESHLY GRATED

1 t MIXED SPICE

400 g COOKED APPLES WELL STRAINED

Mix flour and soda well. Mix in spices. Then blend in a vitamiser the tofu, juice, maple syrup. Add to flour mix and stir well. Keep stirring and develop the mixture for 2 minutes. Add the apples. Bake at 180°C (350°F) for 40 minutes. Home ovens are idiosyncratic, so you really need to judge the right temperature often. Do not make this cake too thick — it ideally suits two cakes which can be joined with a crème.

Substitute date purée for the honey or syrup — an extra ½ cup (i.e. 1½).

ICE BEAN — CHOCOLATE OR CAROB

(a large recipe — halve it if you like)

4 CUPS APPLE JUICE

3 T AGAR FLAKES

4 CUPS SOYMILK

1 T ARROWROOT

3 CUPS DATE PURÉE (SEE *BELOW*)

1 CUP MAPLE SYRUP

2 T VANILLA

3 CUPS ROLLED OATS COOKED

2 T TAHINA

6 T PURE CHOCOLATE POWDER OR CAROB

Cook the rolled oats well as a porridge with water and cool — it should not be thin. Make the date purée by cooking 2 cups dates with enough water to barely cover them. Bring to the boil and simmer until soft. Cool and purée. Dissolve agar flakes in apple juice and bring to the boil. Simmer until agar is dissolved. Keep on a very low simmer. Bring 3½ cups soymilk to the boil. Dissolve arrowroot in remaining ½ cup and add to soymilk. Stir well until slightly thick. Mix chocolate and maple syrup with a whisk until all the chocolate is incorporated. Mix chocolate syrup with dates. Stir soymilk into agar and add chocolate mix. Purée and chill till about to set. Mix rolled oats, tahina, vanilla and purée until very smooth. Add the almost-set gel and purée thoroughly. Freeze. When almost set, purée again and re-freeze.

BIBLIOGRAPHY

Ballentine, Rudolph, *Diet & Nutrition*, Himalayan International Institute, Pennsylvania 1978

Brissenden, Rosemary, *Asia's Undiscovered Cuisine*, Pantheon, New York 1982

Bugialli, Guilano, *The Taste of Italy* Conran Octopus., London 1985

Chang, K.C., (ed.) *Food in Chinese Culture*, Yale University Press, 1977

Davidson, Alan, *Fish & Fish Dishes of Laos*, Tuttle, Tokyo 1975

Downes, John, *Natural Tucker*, Hyland House, Melbourne 1978

Downes, John, *The Natural Tucker Bread Book*, Hyland House, Melbourne 1963

Kiang, Mi Mi, *Cook and Entertain the Burmese Way*, Daw Ma Ma Khin, Rangoon 1975

Kobayashi, K., *Shojin Cooking – the Buddhist Vegetarian Cook Book*, Buddhist Bookstore, Japan 1977

Kushi, Aveline *How to Cook with Miso*, Japan Publications, Tokyo 1978

Sacharoff, Shanta Nimbark, *Flavours of India*, 101 Productions, San Francisco 1972

Shurtleff, W., & Aoyagi, A.,

    The Book of Tofu (vol. 1), Autumn Press, Tokyo 1975

    The Book of Tofu (vol. 2: Tofu and Soy-milk Production), New Age Foods Study Center, California 1979

    The Book of Miso, Autumn Press, Tokyo 1976 (vol. 1)

    The Book of Miso, (vol. 2: Miso Production), New Age Foods Study Center, California 1970

    The Book of Tempeh, Harper & Row, New York 1979

    The Book of Tempeh, (vol. 2: Tempeh Production), New Age Foods Study Center, California 1980

Sonakui, Sibpan, *Everyday Siamese Dishes*, Pracandra Press, Bangkok 1974

Soycrafters Association of North America, *The Beanfield*, newsletters, various editions, and *Soyfoods in America*, annual conference notes on producing and marketing soyfood, The Soyfoods Center, Lafayette, California

Tsuji, Shizuo, *Japanese Cooking – A Simple Art*, Kodansha, Tokyo 1980

Wolf, Roinhart, Tiger, Lionel, & Yin-Fei Lo, Eileen, *China's Food*, Friendly Press, New York 1976